GERMAN ACES OF WORLD WAR I

GERMAN ACES OF WORLD WAR I

THE PICTORIAL RECORD

Norman Franks & Greg VanWyngarden

Schiffer Military History
Atglen, PA

Book design by Robert Biondi.

Library of Congress Catalog Number: 2004105476.

Printed in China.
ISBN: 0-7643-2117-X

We are always looking for people to write books on new and related subjects. If you have an idea for a book, please contact us at the address below.

Published by Schiffer Publishing Ltd.
4880 Lower Valley Road
Atglen, PA 19310
Phone: (610) 593-1777
FAX: (610) 593-2002
E-mail: Info@schifferbooks.com.
Visit our web site at: www.schifferbooks.com
Please write for a free catalog.
This book may be purchased from the publisher.
Please include $3.95 postage.
Try your bookstore first.

In Europe, Schiffer books are distributed by:
Bushwood Books
6 Marksbury Ave.
Kew Gardens
Surrey TW9 4JF
England
Phone: 44 (0)20 8392-8585
FAX: 44 (0)20 8392-9876
E-mail: Bushwd@aol.com.
Free postage in the UK. Europe: air mail at cost.
Try your bookstore first.

INTRODUCTION

Recording the list of accepted German ace pilots and observers of the First World War is not new, but this is the first attempt at producing a book of portraits and photographs of the over 390 aces.

This in itself was a daunting prospect. It was first envisaged as just a collection of portraits of aces with ten or more victories, which was the standard the Germans themselves utilized. In the First World War the Germans did not use the Allied term 'ace', but they did eventually recognize successful fighter pilots as *Kanonen* (literally 'cannons', but perhaps 'top guns' is a better translation in today's parlance). The criterion for this distinction was generally ten aerial victories or more. As the authors collected the photographs it became apparent that to ignore those aces who scored between nine and five victories was to do the project an injustice, so we moved on to assemble as many as we possibly could.

It has often been said that historians and enthusiasts should resist the urge to compare the score of one ace with another, and by the same token they should certainly not fall into the trap of assuming that a score was the only yardstick of success. While some pilots did achieve high scores well in excess of twenty, thirty, or even forty victories, these may have been achieved in less dangerous war operations than others, or while flying aircraft that were far superior to those in use by the Allies at a particular place or time. Not all actions were over the Western Front, for several German aces flew and fought in northern Italy, Macedonia, Palestine and Salonika.

In the early stages of aerial combat, the achievements of pioneering Fokker *Eindecker* (monoplane) pilots like Max Immelmann, Oswald Boelcke, or Kurt Wintgens resulted in great adulation by both their military superiors and the general German public. A pilot who gained eight victories or so could receive high honors like the *Pour le Mérite*. As the war progressed, the intensity and scale of aerial warfare increased dramatically and later pilots had to gain more and more victories to earn such recognition.

In the early days pilots were teaching themselves to become fighter pilots at a time when there were few aircraft in the sky. By mid-1916 the air war was becoming far busier, and that summer saw the introduction of the first '*Jagdstaffeln*' (or *Jastas*): true dedicated fighter units with a strength of fourteen aircraft each. April 1917, or 'Bloody April' as the Royal Flying Corps later called it, saw these emerging *Jagdstaffeln* operating exceptionally good Halberstadt and Albatros fighters, the latter equipped with two synchronized machine guns firing through the arc of the propeller blades. The *Jasta* pilots were tremendously keen to engage an enemy that was not so well equipped. In the race for superior aircraft the Allied nations later developed better machines than those of the German air service: Sopwith Camels, SE5s, Spads and the Bristol F2B two-seaters. However, the balance would shift back and forth as succeeding generations of fighters were produced by both sides. For a time the Fokker Dr.I (triplane) was particularly feared and later the Fokker D.VII biplanes of 1918 gave Allied airmen much to think about, although by then the quality of German pilots was starting to suffer as attrition took its toll and Germany's manpower reserves had been stretched to their limit.

There should not be too much comparison made between some of the early aces, such as Immelmann, Max von Mulzer, and Otto Parschau (who were not only learning how to fight in the air but developing tactics to try and stay alive) and later aces of the 1917-1918 era. The seemingly meager scores of the early aces of 1915-1916 were quickly surpassed by the aces of 1917, who had better equipment, better armament, and the knowledge passed down by the earlier pilots. They were also fighting in what we might now term a target-rich environment.

In all air forces there were men who could and did fly and stay alive for long periods in front-line fighter units but who were unable to score more than one or two victories. The mere fact that they survived the daily aerial combats shows that they were certainly good pilots, but just may not have had the skills to bring down opponents successfully. These men should not be ignored, for often it was the mere fact that they were supporting the more successful pilots (those who had the ability to judge angles of deflection, the fearlessness to get in close before firing, and who were able to stalk an enemy without being seen) that allowed those skilled pilots to shoot down aircraft. This was especially true of the last eighteen months of the conflict, which by then had become a patrol leader's war and no longer an individual one.

The practice of the lone pilot out hunting for trouble in the early days only came about because of the relatively few aircraft in action, and because it was difficult for large formations of machines to operate together. This changed as time passed and with more aircraft in operation and more chances of meeting opposition, the day of the lone hunter had basically passed.

In these flight-sized, sometimes squadron-sized patrols, it was generally the leader who had the best chance to score a victory before the whole affair degenerated into a whirling 'dogfight'. The Germans particularly honed this skill to a fine art, and men such as Manfred von Richthofen developed *Jasta* tactics so that the leader of the attack was able to concentrate on that first clash with the enemy, knowing that his back was protected by his men. In this way the leader was better able to direct his attention to the target and to shoot down an enemy aircraft before the attack developed into a melee. With aircraft twisting and turning, diving and climbing, it then became more of a matter of luck if one pilot was able to stay on the tail of an opponent long enough to get in a telling burst of gunfire, without having someone slip in behind him to do the same.

For much of the time the German fighter pilots, with their defensive stance, had many advantages over the Allied airmen. They did not need to patrol as much as their opponents did. Very often, on clear days, the German pilots could actually observe the sky above the trenches from their airfields that were sometimes only a few miles away. Certainly they could make out hostile aircraft through binoculars or telescopes and then decide if and when to take off and engage them. Otherwise a telephone call from a forward observation post would give the *Jasta* pilots a reasonably accurate picture of where the enemy might be, how many of them there were and what height and direction they were flying. Then again, there were always Allied Corps aircraft doing some work over the lines, directing artillery fire, photographing or bombing. However, once these pilots were in the air there were no radio aids to direct them as in later wars.

Over France there was generally a prevailing westerly or southwesterly wind which helped the Allied airmen push on into German-held territory, but which was against them as they later headed back to their airfields. This too was an advantage for the German fighter pilots, for they could pounce on a struggling enemy whose fuel tanks were starting to run dry and who were heading into something of a head wind. In those moments Allied airmen were not overly keen to see German fighters either coming for them or barring their path home.

Victories by all the warring forces had to be confirmed and each nation had its own methods of claiming aerial victories. In the 1915 era the German Fokker *Eindecker* pilots were required to fight a more defensive war, and were not encouraged to cross the trench lines into Allied air space. This was done in order to keep the technical details of the synchronized machine guns secret, and also because these fighting aircraft were thinly spread over the Front. As the German airmen continued to be outnumbered by their opponents in most sectors, this general defensive tactic remained in place for most of the war. Thus if an Allied aircraft was brought down, it often fell inside German lines so that there was generally a machine or wreckage that could be identified and credited to the victor. Over fluid battle zones this was not so straightforward, nor was it easy to get confirmation if an opponent was seen to 'fall' beyond the lines in Allied territory. In October 1916 the German air service was reorganized, and the office of *Kogenluft,* or the *Kommandieren General der Luftstreitkräfte* (general in command of the German Army Air Service), was created. One of the new offices established was given the responsibility and power to accept, review and decide on the confirmation of aerial combat claims. A *Jasta* pilot would usually submit his victory claim to his commanding officer, who could then choose to send it on to the *Kommandeur der Flieger* or *Kofl* (the officer in charge of all the flying units assigned to his particular German Army). If accepted the *Kofl* sent the claim on to *Kogenluft* for the final decision. It is a fact that in these chaotic air fights, more than one pilot could claim the same victory, or a fighter pilot and a flak unit might both claim to have brought down a particular aircraft. If the conflicting claims occurred within the same *Jasta*, then the dispute might be settled informally (sometimes even by a roll of the dice). If not, then an Arbitration Committee (*Schiedsgericht)* of impartial officers would be convened to decide on whose victory claim was valid, and this claim would then be submitted to the *Kofl.*

The testimony from eyewitnesses, either in the air or on the ground, was eagerly sought by the claimants, but their testimony was only as good as what they thought they saw. There were many examples of claims being made of 'crashes' in which aircraft did not crash and 'burning' aircraft that did not burn (leaking vaporized fuel was often seen as smoke that was either masking or preceding flames). Nevertheless, it must be said that German victory credits were a good deal more accurate than the British claims. As much of their fighting was done over the German lines, British airmen had far fewer wrecks of hostile aircraft to consider. An aircraft seen spinning down after a fight, but not actually seen to crash or burn, was claimed as 'out of control' which in effect was a 'probable' and sometimes only a possible! Usually the enemy pilot was merely spinning down to get out of trouble. Similar German claims were recorded as *zur Landung gewesen,* or 'forced to land', but this did not count as a true aerial victory.

Another advantage German airmen had in 1918 was the availability of parachutes, though their use was not obligatory at first. A number of German pilots were saved as they jumped from burning or crippled machines, and among the photographs in this book one will see pilots with the 'Heinecke' parachute harnesses fixed firmly about them. These new devices did not always work, but at least the pilots had a chance of survival where hitherto they had had almost none in a flaming or disintegrating aircraft. Both sides provided parachutes for their kite balloon observers, but sadly the crews of Allied airplanes were denied the use of this life saving apparatus.

Among the aces were a handful of observers who scored five or more victories from the back seat of a two-seater. More than one German fighter pilot began his scoring as an observer or pilot of a two-seater. These airmen are also represented in the record.

In some cases, the total number of confirmed victories credited to a particular pilot scored is still uncertain or in some dispute by aviation historians. This is particularly true in regard to victories achieved in the last chaotic weeks of the war, when confirmation processing was difficult at best, and for which surviving records are fragmentary. One example is the score of Josef Jacobs; late in his life he himself claimed he had a total of forty-seven victories, though some historians credit him with forty-eight while *Kogenluft* confirmations for only forty-one have been located. When the authors are aware that there is some disagreement or uncertainty about a pilot's score, it has been noted in the photograph captions.

An alphabetical ace listing is provided as an index. Therefore if the score of a particular airman is not know his entry can easily be found by checking through this alphabetical listing.

The photographs are of necessity of varying quality. They came from a wide variety of sources, and many are merely snapshots rather than studio portraits. Some of them were taken on wartime airfields or in front line living quarters, and so we are fortunate they have survived. Many images of the early or high-scoring aces were preserved in the beautiful postcard photographs of Willi Sanke and others, but in regard to the lesser-known aces we are not so fortunate. It was a conscious decision to include all the necessary photos that we could find, working on the basis that a poor photo was better than none at all. In several cases, the only available image was part of a larger group photo. Of those aces who are missing, we hope some will turn up gradually after publication so that there will be the possibility of including them in a future reprint edition.

Any modern historian or enthusiast who collects photographs or information about these early airmen owes a great debt to a small band of amateur German historians who assembled and preserved a good deal of material in the 1920s and 1930s. With dedication and perseverance, collectors like Theo Melchers, Fritz Schmidt, Ludwig Himmelstoss, and Hans Isemann preserved many photographs for later generations, and their names should not be forgotten. Writers and historians such as Werner von Langsdorff and Erich Tornuss also saved much valuable information from oblivion. It is well known that many German archives, both public and private, were destroyed in World War II, and afterwards it was up to a new generation of aviation historians from Germany and abroad to unearth and preserve the material that survived. Many of these enthusiasts (including the authors) benefited from the work and generosity of the late Heinz Nowarra, who started his collection anew after the War. The authors of the present volume built their files with the valuable help of the late Neal W. O'Connor, A.E. "Ed" Ferko, William Puglisi, Dr. Gustav Bock, Tony Svenson, Dr. James Parks, and Bill Evans – all of whom are sorely missed. All enthusiasts of this subject owe a great deal to that indefatigable researcher Alex Imrie, who contacted many surviving German pilots and preserved their photographs; many of the best photos in this book originated with Mr. Imrie in one way or another. We would also like to extend our thanks to our friends Peter M. Grosz, Peter Kilduff, George H. Williams, Robert Gill, Stephen Lawson, Hugh Wynne, Paul S. Leaman, Dr. Volker Koos, Mike O'Connor, Volker Haeusler, Wouter Warmoes, and Reinhard Kastner. The encyclopedic knowledge of researchers such as Rick Duiven, Gunnar Söderbaum, and the respected Manfred Thiemeyer was of great utility and is heartily appreciated. Paul A. Oelkrug, curator of special collections of the McDermott Library of the University of Texas, Dallas, was extremely helpful, as were secretary Carole Thomas and our good friend, library volunteer Ken Rice.

Abbreviations

DoI	Died of Injuries
DoW	Died of Wounds
FA	*Flieger Abteilung*
FFA	*Feld Flieger Abteilung*
Fr.	*Freiherr*
Gfr.	*Gefreiter*
Hptm.	*Hauptmann*
JG	*Jagdgeschwader*
JGr	*Jagdgruppe*
Jasta	*Jagdstaffel*
Kasta	*Kampfstaffel*
Kek	*Kampfeinsitzer Kommando*
Kest	*Kampfeinsitzer Staffel*
KIA	Killed in Action
KIFA	Killed in Flying Accident
KOAS	Killed on Active Service
Ltn.	*Leutnant*
Ltn.z.S.	*Leutnant zur See*
MA.	*Matrosenartillerie*
MI	*Marineinfanterie*
Oblt.	*Oberleutnant*
Off.Stv.	*Offizierstellvertreter*
Rittm.	*Rittmeister*
S/SS	*Schutzstaffel/Schlachtstaffel*

THE PICTORIAL RECORD

***Rittmeister* Manfred Albrecht *Freiherr* von Richthofen**
Pour le Mérite
Born Breslau, 2 May 1892. KIA France, 21 April 1918.
Jastas 2, 11, JG Nr.I
80 victories

***Oberleutnant* Ernst Udet**
Pour le Mérite
Born Frankfurt, 26 April 1896. Died Berlin, 17 November 1941
Kek Habsheim, Jastas 15, 37, 11 and 4
62 victories

***Oberleutnant* Erich Löwenhardt**
Pour le Mérite
Born Breslau, 7 April 1897. KIA France, 10 August 1918
Jasta 10
54 victories

***Leutnant* Josef Carl Peter Jacobs**
Pour le Mérite
Born Kreuzkapelle, Rhineland, 15 May 1894. Died Munich, 29 July 1978
Jastas 12, 22 and 7
48 victories

Jacobs' score is also recorded as 47 or 41. He is seen with one of his famous all-black Fokker triplanes of *Jasta* 7, powered by a captured Clerget engine.

***Leutnant* Werner Voss**
Pour le Mérite
Born Krefeld, 13 April 1897. KIA France, 23 September 1917
Jastas Boelcke, 5, 14, 29 and 10
48 victories

***Leutnant* Fritz Rumey**
Pour le Mérite
Born Königsberg, 3 March 1891. KIA France, 27 September 1918
Jastas 2 and 5
45 victories

Hauptmann **Rudolf Berthold**
Pour le Mérite
Born Ditterswind, 24 March 1891. KOAS Harburg, 15 March 1920.
Kek Vaux, Jastas 4, 14 and 18. JG Nr.II
44 victories

Hauptmann **Bruno Loerzer**
Pour le Mérite
Born Berlin, 22 January 1891. Died 23 August 1960.
Kek Jametz, Jastas 17 and 26, JG Nr.III
44 victories

Leutnant **Paul Wilhelm Bäumer**
Pour le Mérite
Born Duisberg-Ruhrart, 11 May 1896. KIFA nr. Hamburg, 15 July 1927.
Jastas Boelcke and 5
43 victories

***Hauptmann* Oswald Boelcke**
Pour le Mérite
Born Giebichstein, Saxony 19 May 1891. KIA France, 28 October 1916.
FFA 13 and 62, Jasta 2
40 victories
Boelcke assumes a familiar pose in front of his Fokker D.II 352/16 at *Jasta* 2 in September 1916.

Leutnant **Franz Büchner**
Pour le Mérite
Born Leipzig, 2 January 1898. KOAS nr. Leipzig, 18 March 1920.
FFA 270, Jastas 9 and 13
40 victories

***Oberleutnant* Lothar Siegfried *Freiherr* von Richthofen**
Pour le Mérite
Born Breslau, 27 September 1894. KIFA nr. Hamburg, 4 July 1922,
KG 4, Jasta 11
40 victories

***Leutnant* Heinrich Gontermann**
Pour le Mérite
Born Siegen, Westphalia, 25 February 1896. KIFA France, 30 October 1917.
Jastas 5 and 15
39 victories

***Oberleutnant* Carl Menckhoff**
Pour le Mérite
Born Herford, Westphalia, 14 April 1883. Died Switzerland in 1948.
Jastas 3 and 72
39 victories

***Oberleutnant* Karl Bolle**
Pour le Mérite
Born Berlin 20, June 1893. Died Berlin, 9 October 1955.
Jastas 28w and Boelcke
36 victories

***Leutnant* Julius Buckler**
Pour le Mérite
Born Mainz, 28 March 1894. Died Berlin, 23 May 1960.
Jasta 17
36 victories

***Leutnant* Max *Ritter* von Müller**
Pour le Mérite
Born Rottenburg, Bavaria 1 January 1887. KIA France, 9 January 1918.
Jastas Boelcke, 28w and Boelcke
36 victories

***Leutnant* Gustav Dörr**
Born 5 October 1887. Killed 11 December 1928.
Jasta 45
35 victories

***Leutnant* Otto Könnecke**
Pour le Mérite
Born Strasbourg, 20 September 1892. Died 25 January 1956.
Jastas 25 and 5
35 victories

***Hauptmann* Eduard *Ritter* von Schleich**
Pour le Mérite
Born Munich, 9 August 1888. Died as British PoW, 15 November 1947.
Jasta 21s and JGr.8
35 victories
Schleich is seen with the black-fuselaged Albatros D.Va he flew as commander of Bavarian *JGr.8*.

***Leutnant* Emil Thuy**
Pour le Mérite
Born Hagan, Westphalia, 11 March 1894. KIFA Smolensk, Russia, 11 June 1930.
Jastas 21s and 28w
35 victories

***Leutnant* Josef Veltjens**
Pour le Mérite
Born Geldern, nr. Duisberg, 2 June 1894. KOAS Yugoslavia, 6 October 1943.
Jastas 14, 18 and 15
35 victories

***Leutnant* Heinrich Bongartz**
Pour le Mérite
Born Gelsenkirchen, 31 January 1892. Died Rheinberg, 23 January 1946.
Jasta 36
33 victories

***Oberleutnant* Heinrich Claudius Kroll**
Pour le Mérite
Born Flatsby, Flensberg, 3 November 1894. Died Hamburg, 21 February 1930.
Jastas 9 and 24s
33 victories

***Oberleutnant* Kurt Wolff**
Pour le Mérite
Born Greifswald, Pomerania, 6 February 1895. KIA France, 15 September 1917.
Jastas 11, 29 and 11
33 victories

***Leutnant* Hermann Frommherz**
Born Waldshut, Baden 10 August 1891. Died Waldshut, 30 December 1964.
Jastas Boelcke and 27
32 victories

***Leutnant der MA* Theodor Osterkamp**
Pour le Mérite
Eorn Aschersleben, nr. Duren, Rhineland, 15 April 1892.
Died Baden-Baden, 2 January 1975.
Marine Feld Jastas I and II
32 victories (or 31)
'Theo' Osterkamp achieved further victories in World War II. He is seen with Fokker E.V 156/18, which he flew for only a very brief time.

***Leutnant* Paul Billik**
Born Haatsch, Silesia, 27 March 1881. KIFA Berlin, 8 March 1926.
Jastas 12, 7 and 52
31 victories

***Oberleutnant zur See* Gotthard Sachsenberg**
Pour le Mérite
Born Dessau, 6 December 1891. Died 23 August 1961.
Marine Feld Jasta I, Marine Feld JGr.I and Marine Feld JG
31 victories

***Leutnant* Karl Allmenröder**
Pour le Mérite
Born Wald, Rhineland, 3 May 1896. KIA France, 27 June 1917
Jasta 11
30 victories

***Leutnant* Carl Degelow**
Pour le Mérite
Born Münsterdorf, 5 January 1891. Died Hamburg, 9 November 1970.
Jastas 7 and 40s
30 victories

***Leutnant* Josef Mai**
Born 3 March 1887. Died 18 January 1982.
Jasta 5
30 victories

***Leutnant* Ulrich Neckel**
Pour le Mérite
Born 23 January 1898. Died Italy, 11 May 1928.
Jastas 12, 13, 19 and 6
30 victories

***Leutnant* Karl Emil Schäfer**
Pour le Mérite
Born Krefeld, 17 December 1891. KIA France, 5 June 1917.
Jasta 11
30 victories

Oberleutnant **Harald Auffarth**
Died 12 October 1946.
Jastas 18 and 29
29 victories
The aircraft behind Auffarth is a Pfalz D.VIII of *Jasta 29*.

Leutnant **Walter von Bülow-Bothkamp**
Pour le Mérite
Born Borby, Holstein, 24 April 1894. KIA France, 6 January 1918.
Jastas 18, 36 and Boelcke
28 victories
Von Bülow poses between two RFC men, the crew of an FE2d he shot down on 7 May 1917 – his 13th victory.

Leutnant **Walter Blume**
Pour le Mérite
Born Hirschberg, Silesia, 10 January 1896. Died 27 May 1964.
Jastas 26 and 9
28 victories

***Oberleutnant* Robert *Ritter* von Greim**
Pour le Mérite
Born Bayreuth, Bavaria, 22 June 1892. Died Saltzburg, 24 May 1945 (suicide).
Jasta 34b, JGr.10 and JGr.9
28 victories

***Leutnant* Arthur Laumann**
Pour le Mérite
Born Essen, 4 July 1894. Died Münster, 18 November 1970.
Jastas 66 and 10
28 victories

***Oberleutnant* Friedrich *Ritter* von Röth**
Pour le Mérite
Born Nürnberg, 29 September 1893. Died Nürnberg, 31 December 1918 (suicide).
Jastas 23b and 16b
28 victories

***Oberleutnant* Fritz Otto Bernert**
Pour le Mérite
Born Ratibor, Upper Silesia, 6 March 1893. Died 18 October 1918.
Jastas 4, Boelcke and 6
27 victories

***Leutnant* Otto Frühner**
Born Brieg, Silesia, 6 September 1893. Died Villach, Austria, 19 June 1965.
Jasta 26
27 victories

Leutnant **Hans Kirschstein**
Pour le Mérite
Born Koblenz, 5 August 1896. KIFA France, 16 July 1918.
Jasta 6
27 victories

Leutnant **Karl Thom**
Pour le Mérite
Born Freystadt, Prussia, 19 May 1893. Missing, Russia, World War II.
Jasta 21s
27 victories

***Hauptmann* Adolf *Ritter* von Tutschek**
Pour le Mérite
Born Ingolstadt, Bavaria, 16 May 1891. DoW France, 15 March 1918.
Jasta 2 and 12, JG Nr.II
27 victories

***Leutnant* Kurt Erwin Wüsthoff**
Pour le Mérite
Born Aachen, 27 January 1897. KIFA Dresden, 18 July 1926.
Jasta 4
27 victories

Oberleutnant **Oskar *Freiherr* von Boenigk**
Pour le Mérite
Born Siegersdforf, Silesia, 25 August 1893. Died Russia, 30 January 1946.
Jasta 4 and 21s, JG.II
26 victories

Oberleutnant **Eduard *Ritter* von Dostler**
Pour le Mérite
Born Pottenstein, Bavaria, 3 February 1892. KIA France, 21 August 1917.
Jastas 13, 34b, and 6
26 victories
The aircraft is an Albatros D.III

***Leutnant* Max Näther**
Born Tepliwada, Prussia, 24 August 1899. KOAS Kolmar, 8 January 1919.
Jasta 62
26 victories

***Leutnant* Oliver *Freiherr* von Beaulieu-Marconnay**
Pour le Mérite
Born Berlin-Charlottenburg, 14 September 1898. DoW France, 26 October 1918.
Jastas 15 and 19
25 victories

***Leutnant* Fritz Pütter**
Pour le Mérite
Born Duelmen, Westphalia, 14 January 1895. DoI Bonn, 10 August 1918.
Jastas 9 and 68
25 victories

***Leutnant* Georg von Hantelmann**
Born Rokietnice, 9 October 1898. Died Baborowko, E. Prussia, 7 September 1924 (murdered by poachers on his estate).
Jasta 15
25 victories
The aircraft is a Fokker D.VII; photo taken at Charmois near Stenay, on 9 October 1918 following his 20th victory.

***Leutnant* Erwin Böhme**
Pour le Mérite
Born Holzminden, 29 July 1879. KIA France, 29 November 1917.
Jastas 2 and 29
24 victories

Böhme is on the right, with his observer *Ltn.* Lademacher at left, posing with their dragon-decorated Albatros C.III C.766/16 which they flew in *Kasta 10* of *Kagohl 2* on the Russian Front, August 1916.

***Leutnant* Georg Meyer**
Born Bremen, 11 January 1893. Died 15 September 1926 (motorcycle accident).
Jastas 22s, 7 and 37
24 victories

***Leutnant* Hermann Becker**
Born 10 September 1887.
Jasta 12
23 victories

***Oberleutnant* Hermann Wilhelm Göring**
Pour le Mérite
Born Rosenheim, Bavaria, 12 January 1893. Died Nürnberg, 15 October 1946 (suicide).
Jastas 26 and 27, JG.I
22 victories

***Oberleutnant* Hans Klein**
Pour le Mérite
Born Stettin, 17 January 1891. Died 18 November 1944.
Jastas 4 and 10
22 victories

Klein is second from left in this group of *JG I* dignitaries and others at the Pfalz aircraft factory in Speyer, with Pfalz D.XII 2603/18. From l. to r.: *Oblt.* Seibert, *Oblt.* Klein, Pfalz factory owner Alfred Eversbusch, *Ltn.* Ernst Udet (62), *JG I* technical officer *Ltn.* Konstantin Krefft (2), Austrian *Oblt.* Lucas, and Speyer Bürgermeister Dr. Moericke. Note Klein's missing right index finger, lost when he was wounded on 19 February 1918.

***Leutnant* Hans Martin Pippart**
Born Mannheim, Baden, 14 May 1888. KIA France, 11 August 1918.
Kasta 1, Jastas 13 and 19
22 victories

Leutnant **Werner Preuss**
Born 21 September 1894. KIFA 6 March 1919.
Jasta 66
22 victories
(Despite the retouched photo, Preuss did not receive the Pour le Mérite.)

Vizefeldwebel **Karl Paul Schlegel**
Born Wechselburg, Saxony, 7 May 1893. KIA France, 27 October 1918.
Jasta 45
22 victories

***Leutnant* Rudolf Windisch**
Pour le Mérite
Born Dresden, 27 January 1897. Killed France, c.27 May 1918.
Jastas 32 and 66
22 victories

***Leutnant* Hans *Ritter* von Adam**
Born Eisenstein, Bavaria, 24 May 1886. KIA France, 15 November 1917.
Jastas 34b and 6
21 victories

Offizierstellvertreter **Friedrich Altemeier**
Born Niederbecksen, nr. Hannover, 14 June 1886.
Jasta 24s
21 victories

Leutnant **Friedrich 'Fritz' Friedrichs**
Born Spark, Westphalia, 21 February 1895. KIA France, 15 July 1918.
Jasta 10
21 victories
The aircraft is a Pfalz D.III of *Jasta 10.*

Leutnant **Fritz Höhn**
Born Wiesbaden, 31 May 1896. KIA France, 3 October 1918.
Jastas 21s, 81, 60 and 41
21 victories

Leutnant **Friedrich Theodor Noltenius**
Born Bremen, 8 January 1894. KIFA Johannisthal, 12 March 1936.
Jastas 27, 6 and 11
21 victories

Leutnant **Karl Odebrett**
Born Schneidemuhl, 31 July 1890. Died Venezuela, 13 February 1930.
Jastas 16 and 42
21 victories

Oberleutnant **Hans Bethge**
Born Berlin 6, December 1890. KIA France, 17 March 1918.
Jastas 1 and 30
20 victories

Leutnant **Rudolf von Eschwege**
Born Bad Homburg v.d. Hole, 25 February 1895. KIA Macedonia, 21 November 1917
Flieger-Abteilungen 66 and 30
20 victories
Eschwege is examining the burned out wreckage of his Halberstadt fighter on Drama airfield, Macedonia.

Leutnant **Wilhelm Frankl**
Pour le Mérite
Born Hamburg, 20 December 1893. KIA France, 8 April 1917.
FA 40 and Jasta 4
20 victories

Leutnant **Hans von Freden**
Born Berlin 18 March 1892. Died 30 October 1919 (Influenza pandemic).
Jastas 1 and 50
20 victories
The aircraft is an Albatros D.Va of *Jasta 1*.

Leutnant **Walter Göttsch**
Born Altour, Hamburg, 10 June 1896. KIA France, 10 April 1918.
Jastas 8 and 19
20 victories

***Vizefeldwebel* Oskar Hennrich**
Jasta 46
20 victories
Hennrich poses with his Fokker D.VII of *Jasta 46.*

***Oberleutnant* Otto Kissenberth**
Pour le Mérite
Born Landshut, Bavaria, 26 February 1893. Died 2 August 1919 (climbing accident in the Bavarian Alps).
Kek Ensisheim, Jastas 16b and 23b
20 victories

***Hauptmann* Wilhelm Reinhard**
Born Düsseldorf, 12 March 1891. KIFA Adlershof, 3 July 1918.
Jastas 11 and 6, JG Nr.I
20 victories

***Vizefeldwebel* Christian Friedrich Donhauser**
KIFA Koblenz, 1919.
Jasta 17
19 victories

***Leutnant* Gerhard Fieseler**
Born Bonn/Rhein, 15 April 1896. Died Kassel, 1 September 1987.
Jasta 25
19 victories

***Oberleutnant* Otto Schmidt**
Born Neukirchen-Saar, 23 March 1885.
Jastas 7, 32b, 29 and 5
19 victories

***Leutnant* Kurt Wintgens**
Pour le Mérite
Born Neustadt, 1 August 1894. KIA France, 25 September 1916.
FA 6, 46 and 6, Kek Vaux, Jasta 1
19 victories

***Leutnant* Hartmut Baldamus**
Born Dresden, 10 August 1891. KIA France, 14 April 1917.
FA 20 and Jasta 9
18 victories

***Leutnant* Franz Hemer**
Born 6 June 1894. Died Frankfurt, 18 October 1982.
Jasta 6
18 victories

***Leutnant* Ernst Hess**
Born Wiesbaden, 8 January 1893. KIA France, 23 December 1917.
Kek Douai, Jastas 28w and 19
17 victories

***Leutnant* Rudolf Klimke**
Born Merseberg, 8 November 1890.
Jasta 27
17 victories

***Leutnant* Karl Plauth**
Born Munich, 27 August 1896. KIFA 1 November 1927.
Jastas 20 and 51
17 victories

***Leutnant* Franz Ray**
Jasta 1, 28w and 49
17 victories
Ray sits in his Albatros D.III of *Jasta 28w.*

***Leutnant* Hans Joachim Rolfes**
Born Port Elizabeth, South Africa, 18 April 1894. KIFA Johannisthal, 12 August 1935
Jastas 32b and 45
17 victories

***Vizefeldwebel* Josef Schwendemann**
Jasta 41
17 victories
The photo is of *Jasta 41* pilots. From l. to r: *Ltn.* Josef Strauch (1), *Ltn.* Georg Schlenker (14), *Ltn.* Ulmer, *Oblt.* Bilke (*OzbV*), Josef Schwendemann, *Vfw.* Iversen, Hermann Juncke (5), Hans Weiss (16), and Julius Reimann (1).

***Leutnant* Ernst Bormann**
Born Kirchbreck, nr. Holzminden, 3 November 1897. Died Düsseldorf, 1 August 1960.
Jasta Boelcke
16 victories

***Leutnant* Ludwig Hanstein**
Born Sudwalde, Prussia, 20 January 1892. KIA France, 21 March 1918.
Jasta 16b and 35b
16 victories

***Leutnant* Wilhelm Anton Seitz**
Jastas 8 and 68
16 victories

***Leutnant* Hans Weiss**
Born Hof, 19 April 1892. KIA France, 2 May 1918.
Jastas 41, 10 and 11
16 victories

***Vizefeldwebel* Karl Bohnenkamp**
Jasta 22s
15 victories

The photo shows pilots from *Jasta 22* in 1917. From l. to r: Kurt Monnington (8), *Vfw*. Bergmann, *Ltn*. Unger, *Ltn*. Julius Fichter (7), *Ltn*. Söbeck, *Vfw*. Karl Bohnenkamp, *Ltn*. Siegfried Büttner (13), *Oblt*. Heinrich Karbe (2), *Ltn*. Albert Wünsche, Alfred Lenz (6), and Ulrich Fischter (4).

***Leutnant* Albert Dossenbach**
Born St. Blasien, Baden, 5 June 1891. KIA France, 3 July 1917.
FA22, Jasta Boelcke, 36 and 10
15 victories

***Leutnant* Rudolf Francke**
Jasta 8
15 victories

***Vizefeldwebel* Albert Haussmann**
Born Stuttgart, 10 October 1892. KIA France, 16 October 1918.
Jastas 23b, 15 and 13
15 victories

***Leutnant* Aloys Heldmann**
Born Grevenbrück, 2 December 1895. Died Grevenbrück, 1 November 1983.
Jasta 10
15 victories

***Oberleutnant* Max Immelmann**
Born Dresden, 21 September 1890. KIA France, 18 June 19 6.
FFA 62
15 victories

***Leutnant* Johannes Klein**
Jastas 18 and 15
15 victories

***Leutnant* Otto Löffler**
Jasta Boelcke
15 victories
Löffler is seen with his Fokker Dr.I 190/17.

Oberleutnant **Hans-Georg von der Marwitz**
Born Ohlau, Silesia, 7 August 1893. KIFA Breslau, 12 May 1925.
Jasta 30
15 victories
Marwitz is on the right with a 208 Squadron Camel pilot next to him, whom he shot down on 16 May 1918 for his 7th victory. The aircraft is a Pfalz DIIIa.

Offizierstellvertreter **Edmund Nathanael**
Born Dielsdorf, 18 December 1889. KIA France, 11 May 1917.
Jasta 22s and 5
15 victories

Leutnant **Wilhelm Neuenhofen**
Born Münich, Gladbach, 24 April 1897. KIFA Dessau, 24 January 1936.
Jasta 27
15 victories

Leutnant **Viktor von Pressentin** ***genannt*** **von Rautter**
Born Willkam, Königsberg, 8 May 1896. KIA France, 31 May 1918.
Jasta 4
15 victories

***Leutnant* Theodor Quandt**
Born Molland, Prussia. 22 June 1897. KIA France, 6 June 1940 (Flying an Me 109E during the Battle of France).
Jasta 36
15 victories

***Leutnant* Julius Schmidt**
Died 2 July 1944.
Jasta 3 and 6
15 victories

Leutnant **Kurt Schneider**
Born Wurzen, nr. Leipzig, 4 October 1888. DoW France, 14 July 1917.
Jasta 5
15 victories

Leutnant **Paul Strähle**
Born Schorndorff, Württemberg, 20 May 1893. Died in 1985.
Jastas 18 and 57
15 victories

Offizierstellvertreter **Reinhold Jörke**
Jastas 12, 13 and 39
14 victories

Leutnant **Franz Piechulek**
Kest 5, Jastas 41 and 56
14 victories
The picture depicts the pilots of *Jasta 56* at Neuvilly, early 1918. Standing l. to r: *Uffz.* Gullmann (1), *Uffz.* Kaleta (2), *Uffz.* Dannemann (2), *Vfw.* Iversen (1), *Vfw.* Weimar (3), *Ltn.* Franz Piechulek; Seated: *Ltn.* Ludwig Beckmann (7), *Ltn.* Franz Schlieff (13), and *Ltn.* Rudolf Heins (4).

Leutnant **Georg Schlenker**
Jastas 3 and 41
14 victories

***Oberleutnant* Rudolf Wendelmuth**
Born Gotha, 28 July 1890. KIA France, 30 November 1917.
Jastas 8 and 20
14 victories

***Hauptmann* Hans-Joachim Buddecke**
Pour le Mérite
Born Berlin, 22 August 1890. KIA France, 10 March 1918.
FFA 23 and Ottoman FFA 5 and 6, Jastas 4, 30 and 18
13 victories

***Leutnant* Siegfried Büttner**
Jastas 22s and 61
13 victories
In the photo of *Jasta 22*, Büttner is seventh from the left (for all names see photograph under Bohnenkamp on page 60.)

***Oberleutnant zur See* Friedrich Christiansen**
Pour le Mérite
Born Wyk-on-Fohr, 12 December 1879. Died Innien, 5 December 1972.
Seeflug 1, Zeebrugge
13 victories

***Leutnant* Dieter Collin**
Born Luben, 17 February 1893. DoW Belgium, 13 August 1918.
Jastas 2, 22s and 56
13 victories

***Leutnant* Heinrich Georg Geigl**
Born Bad Abbach, Bavaria, 10 July 1891. KIA France, 4 April 1918.
Jastas 34b and 16b
13 victories

***Offizierstellvertreter* Robert Heibert**
Born Oberfall, Mosel, 8 January 1886. Died 10 May 1933 (suicide).
Jastas 33 and 46
13 victories

***Leutnant* Johann Janzen**
Born Fronza, West Prussia, 21 May 1896. Died in the 1980s.
Jastas 23b and 6
13 victories
Janzen stands in front of his *Jasta 6* Fokker Dr.I 403/17.

***Vizefeldwebel* Otto Rosenfeld**
Born Nitleben, 4 May 1892. KIA France, 7 July 1918.
Jastas 12 and 41
13 victories
The aircraft is a *Jasta 12* Albatros D.III.

***Oberflugmeister* Kurt Schönfelder**
Born Totschen, 30 July 1894. KIA, Belgium, 26 June 1918.
Jasta 7
13 victories

The picture shows the pilots of Jasta 7 in November 1917. Standing l. to r: *Vfw*. Swoboda, *Ltn*. Paul Billik (31), *Vfw*. Rudolf Lander, *Ltn*. Paul Lotz (9), *Ltn*. Max Hillmann (1), *Ltn*. Mühlen, *Obflm*. Kurt Schönfelder, *Ltn*. Carl Degelow (30), *Vfw*. Hans Horst; Seated: *Ltn*. Hermann Kunz (6), *Ltn*. Josef Jacobs (48), *Ltn*. Otto Kunst (1).

***Oberleutnant* Erich Rüdiger von Wedel**
Born Neuwedel, 9 March 1892. Died 2 July 1954.
Jasta 11
13 victories

Vizefeldwebel **Erich Buder**
Jastas 5 and 26
12 victories

Oberleutnant **Theodor Cammann**
Jastas Boelcke and 74
12 victories

***Vizefeldwebel* Gottfried Ehmann**
Observer with Schlachtstaffel 15
12 victories
In the picture Ehmann is with his pilot Friedrich Huffzky (9), seated in a Halberstadt CL.II of *Schlachtstaffel 15*.

***Offizierstellvertreter* Otto Esswein**
Born Waiblingen, Württemberg, 3 March 1890. KIA France, 21 Ju y 1918.
Jasta 26
12 victories
The picture shows Fritz Classen (11), Otto Frühner (27) and *Offz. Stv.* Otto E-swein in front of a line-up of *Jasta 26* Fokker Triplanes in April 1918.

***Vizefeldwebel* Sebastian Festner**
Born Holzkirchen, Bavaria, 30 June 1894. KIA France, 25 April 1917.
Jastas 5 and 11
12 victories

***Leutnant* Wilhelm Frickart**
Observer FFA24 and FA 242; pilot with FA 20 and Jasta 64 and 65
12 victories
(Observer to Leopold Anslinger, 10 victories).

***Leutnant der MA* Bertram Heinrich**
Born Charlottenberg, 29 March 1894. KIA Belgium, 31 August 1918.
Marine Feld Jasta I
12 victories (Score listed as 9 or 10 in some sources)
The aircraft is Fokker E.V 160/18.

***Leutnant* Walter Höhndorf**
Pour le Mérite
Born Prutske, 10 November 1892. KIFA France, 5 September 1917.
FFA12, Kek Vaux, Jastas 1, 4, and 14
12 victories

***Flugmeister* Gerhard Hubrich**
Born 30 July 1896. Died 20 October 1972.
Seefrontstaffel, Marine Jasta IV
12 victories

***Leutnant* Hans von Keudell**
Born Berlin, 5 April 1892. KIA France, 15 February 1917.
Jastas 1 and 27
12 victories

***Vizefeldwebel* Max Kuhn**
Died in 1920.
Jasta 21s
12 victories

The photo depicts pilots of Saxon *Jasta 21* circa August 1918. l. to r: *Uffz.* Heinrich Haase (5), *Vfw.* Max Kuhn, *Ltn.* Karl Thom (27), *Ltn.* Josef Seebald, *Oblt.* von Kalkreuth (*OzbV*), *Oblt.* Oskar *Fr.* von Boenigk (26), *Oblt.* von Brockel (1), *Ltn.* Julius Keller (2), *Ltn.* Christians (2), *Ltn.* Hilmar Glöcken.

***Leutnant* Alfred Lindenberger**
Born Stuttgart, 22 April 1897. Died Nürtingen, 30 June 1973.
Observer FA 234, pilot Jasta Boelcke
12 victories

Lindenberger is seated in his *Jasta Boelcke* Fokker D.VII (OAW) 4453/18. He also scored four victories in World War II.

Vizefeldwebel **Friedrich Manschott**
Born Reichartshausen, 21 February 1893. KIA France, 16 March 1917.
FA 203, Jasta 7
12 victories

Leutnant **Hans Müller**
Born Etzel, Friesland, 3 July 1896.
Jastas 12, 15 and 18
12 victories

***Oberleutnant* Franz Schleiff**
Born Altchristburg, Prussia, 19 September 1896.
Jastas 41 and 56
12 victories
Schleiff's left hand was amputated after a wound suffered on 27 March 1918; he sits on a Fokker D.VII here after his injury.

***Offizierstellvertreter* Bernhard Ultsch**
Born Wunseidel, Bavaria, 26 March 1898.
Schutzstaffel 29, Jastas 39 and 77b
12 victories

Leutnant **Richard Wenzl**
Died in 1957.
Kek Ost, Jastas 31, 11 and 6
12 victories

Leutnant **Heinrich Arntzen**
Born Aubolt, 11 September 1894.
FA2, Jastas 15 and 50
11 victories

***Leutnant* Raven *Freiherr* von Barnekow**
Born Schwerin, 10 March 1897. Died 25 October 1942 (suicide following the death of his friend Ernst Udet).
Jastas 4, 11, 20, 1 and 20
11 victories

***Leutnant* Joachim von Busse (centre)**
Born Latkovo, Hohensalzen, 14 May 1893. Missing Isterburg, January 1945.
Jastas 3 and 20
11 victories

***Vizefeldwebel* Fritz Classen**
Jasta 26
11 victories
Photo shows Jasta 26 in May 1918: Standing l. to r: *Vfw*. Erich Buder (12), *Vfw*. Fritz Classen, *Ltn*. Claus Riemer (7), *Ltn*. Fritz Loerzer (11), *Oblt*. Bruno Loerzer (44), *Ltn*. Otto Marquardt (1), *Vfw*. Herbert Werner (1), *Off. Stv*. Otto Esswein (12), *Vfw*. Otto Frühner (27), *Ltn*. Eckert; left window: *Ltn*. Zogmann and *Ltn*. Weiss (*OzbV*), right window: *Ltn*. Werner Dahm (4) and *Off.Stv*. Fritz Kublum (1). (also see photo on page 76)

***Leutnant* Xaver Dannhuber**
Jasta 26
11 victories
Aircraft is an Albatros D.V of *Jasta 26*.

***Rittmeister* Kurt-Bertram von Döring**
Born Ribbekardt, Silesia, 18 September 1889. Died Medingin, 9 July 1960.
Jasta 4, JGr 4, Jastas 66 and 1
11 victories

***Leutnant* Heinrich Drekmann**
Born Hamburg, 5 April 1896. KIA France, 30 July 1918.
Jastas 26 and 4
11 victories

This photo shows pilots of *Jasta 4* in the early summer of 1918. l. to r: *Ltn.* Julius Bender, *Ltn.* Egon Koepsch (9), *Ltn.* Karl 'Carlos' Meyer with back to camera (4), *Ltn.* Heinrich Drekmann, *Ltn.* Heinrich Maushake – reclining and obscured under the wing – (6) and *Ltn.* Ernst Udet (62) standing by the wing. The aircraft are Fokker Triplanes, and on the right is Udet's Dr.I 586/17, striped in black and white.

***Vizefeldwebel* Willy Gabriel**
Born Berlin, 31 December 1893. Died Berlin, 29 April 1968.
SS15, Jasta 11
11 victories
Willy Gabriel is on the right with walking stick, and his gunner *Vfw.* Hilsbein is on the left, with their Halberstadt CL.II of *Schlasta 15.*

***Oberleutnant* Stefan Kirmaier**
Born Lachen, 28 July 1889. KIA France, 22 November 1916.
Kek Metz and Kek Jametz, Jasta 2
11 victories
Kirmaier poses with a Fokker *Eindecker*.

***Leutnant* Fritz Loerzer**
Born 27 July 1893. Died 21 July 1952.
Jastas 6, 26, 63 and 26
11 victories

***Leutnant* Hermann Pfeiffer**
Born Tengen, Baden, 24 July 1890. KIFA France, 20 May 1917 (testing a captured French Nieuport 17).
AOK 3, Jasta 9
11 victories

Leutnant zur See **Reinhold Poss**
KIFA Hagenow, 26 August 1933.
Seefrontstaffel, Marine Feld Jasta IV
11 victories

Leutnant **Hugo Schäfer**
Born Eberfeld, 30 June 1894. KIFA 3 February 1920.
Jastas 18, 15
11 victories
Schäfer sits in his snake-decorated *Jasta 18* Albatros D.V in this photo.

***Leutnant* Rudolf Stark**
Born Bavaria, 11 February 1897.
Jasta 34b, 77b and 35b
11 victories
Stark is seen in his *Jasta 34b* Pfalz D.IIIa.

***Vizefeldwebel* Christian Mesch** (NO PHOTO)
KOAS Baltic, 1919.
Jasta 26
13 victories

***Vizefeldwebel* Hans Nülle** (NO PHOTO)
Born Gross Heiden, 27 December 1897.
Jasta 39
11 victories

***Leutnant* Renatus Theiller** (NO PHOTO)
Born Mulhausen, 13 September 1894. KIA France, 24 March 1917.
Jasta 5
11 victories

***Leutnant* Leopold Anslinger**
Born Freiberg im Breisgau, Baden, 13 August 1891. Died 18 June 1978.
FFA 54, FFA 24 and FA 242
10 victories
Anslinger's aircraft is a Roland D.II.

***Offizierstellvertreter* Paul Aue**
From Saxony. Died Russia, 1945 (as a Russian prisoner of war).
Jasta 10
10 victories
Aue is seen with Fokker D.VII 5136/18 in a post-war *Freikorps* unit.

***Oberleutnant* Hans Berr**
Pour le Mérite
Born Braunschweig, 20 May 1890. KIA France, 6 April 1917.
Kek Avillers, Jasta 5
10 victories

***Leutnant* Otto Brauneck**
Born Sulzbach, 27 February 1896. KIA France, 26 July 1917.
FA 69, Jastas 25 and 11
10 victories

***Leutnant* Martin Demisch**
Born Bautzen, Saxony, 12 November 1896. DoW France, 25 September 1918 (shot down on the 24th).
Jasta 58
10 victories
The aircraft is an Albatros D.Va of *Jasta 58*.

***Leutnant* Karl Gallwitz**
Born Sigmaringen, Hohenzollern, in 1895.
FA (A) 231, FA 37, Jasta Boelcke
10 victories

***Leutnant* Justus Grassmann**
Born 19 January 1897. Died 2 November 1961.
Jasta 10
10 victories
Grassmann sits in the cockpit of his *Jasta 10* Pfalz D.IIIa.

***Leutnant* Paul Lotz**
Born Rumenau, 2 October 1895. KIFA France, 23 October 1918.
Jastas 7 and 44
10 victories
Jasta 7 pilots. Seated l to r: *Ltn*. Hermann Kunz (7), *Ltn*. Josef Jacobs (48), *Ltn*. Otto Kunst (1); standing: *Vfw*. Swoboda, *Ltn*. Paul Billik (31), *Ltn*. Paul Lotz, *Ltn*. Max Hillmann (1), *Ltn*. Mühlen, *Obflgm*. Kurt Schönfelder (13), *Ltn*. Carl Degelow (30) and *Vfw*. Hans Horst (1).

***Leutnant* Rudolf Matthaei**
Born Hildesheim, 10 November 1895. KIFA France, 17 April 1918.
Jastas 21s, 5 and 46
10 victories

***Leutnant* Max *Ritter* von Mulzer**
Pour le Mérite
Born Kamratshofen, Bavaria, 9 July 1893. KIFA France, 26 September 1916.
FFA 62, Kek Vaux, Kek Bertincourt, and FFA 32
10 victories

***Vizefeldwebel* Alfons Nagler**
Born Ertingen, Württemberg, 4 August 1894.
Jasta Ober-Ost, Jastas 74 and 81
10 victories

***Oberleutnant* Hans Schüz**
KIFA 31 August 1941.
Ottoman FFA 6, FFA 2 and FFA 13 (operating over Macedonia, Gallipoli and Palestine)
10 victories
The aircraft is an Albatros D.III with Turkish markings and twin radiators.

***Leutnant* Werner Steinhäuser**
Born Königsberg, 29 June 1893. KIA France, 26 June 1918.
Jasta 11
10 victories

***Leutnant* Erich Thomas**
Jastas 9 and 22s
10 victories

***Leutnant* Paul Turck**
Jastas 66 and 21s
10 victories

***Leutnant* Paul Wenzel**
Born Seemuhl, 24 January 1887. Died Germany, 30 January 1964.
Jasta 41 and 6
10 victories

Leutnant **Hans Joachim Wolff**
Born Mülhausen, 24 September 1895. KIA France, 16 May 1918.
Jasta 11
10 victories

Oberleutnant **Ernst *Freiherr* von Althaus**
Pour le Mérite
Born Coburg, Bavaria, 19 March 1890. Died Berlin, 29 November 1946.
Kek Vaux, Jastas 4, 14 and 10
9 victories

***Leutnant* Arnd Benzler**
Jastas 32b, 45, 65 and 60
9 victories
The aircraft is a Fokker D.VII of *Jasta 60*.

***Leutnant* Albert Dietlen**
Born Heilbronn, Baden, 28 July 1889. KIA France, 12 April 1918.
Jastas 23b, 41 and 58
9 victories

***Leutnant* Otto Fitzner**
Jastas 17 and 65
9 victories
Photo of *Jasta 17* group. From bottom to top: *Ltn.* Ernst Wendler (CO, hands on knees), *Ltn.* Günther Schuster (6) to the right, *Vfw.* Gustav Schniedewind (7) to the left, *Ltn.* Alfred Traeger (3), *Ltn.* Wilhelm Becker (2), *Ltn.* Erich Zschunke (*OzbV*), *Ltn.* Ehlers, *Ltn.* Walter Brachwitz (2), *Vfw.* Julius Buckler (35), *Ltn.* Otto Fitzner (fourth from top), *Ltn.* Jakob Wolff (4), *Vfw.* Adolf Werner (2) and *Vfw.* Georg Strasser (7).

***Feldwebel* Friedrich Huffzky**
Schlachtstaffel 15
9 victories
His observer/gunner, with the rear seat Parabellum gun, is Gottfried Ehmann (12).

***Leutnant* Egon Koepsch**
Born 27 October 1890. Died 26 November 1976.
Jasta 4, 11 and 4
9 victories

***Offizierstellvertreter* Fritz Kosmahl**
Born Leipzig, 5 September 1892. DoW France, 26 September 1917 (mortally wounded 22 September.).
FFA 22, FA 261, Jasta 26
9 victories

***Leutnant* Walter Kypke**
Born Stettin, 22 September 1893. KIFA Magdeburg, 6 June 1924.
Kek Avillers, Jasta 41, Kest 5, Jasta 47w
9 victories

***Leutnant* Helmut Lange**
Jasta 26
9 victories
The photo shows *Jasta 26* on 5 February 1918 after Albert Ball's brother Cyril had been shot down and captured by Otto Esswein. L. to r: *Vfw*. Fritz Classen (10), unknown, *Vfw*. Erich Buder, *Uffz*. Jon Santjer, ground crewman, *Off.Stv*. Otto Esswein (12), unknown, Cyril Ball RFC, *Ltn*. Werner Dahm (2), *Oblt*. Bruno Loerzer (44), *Ltn*. Helmut Lange, *Uffz*. Friedrich Naubauer (1), *Ltn*. Willy Etzold, *Ltn*. von der Burchard. The aircraft is a Fokker D.V used to convert pilots to rotary-engined machines pending the arrival of Fokker Triplanes.

Leutnant **Gustav Leffers**
Pour le Mérite
Born Wilhelmshaven, 2 January 1892. KIA France, 27 December 1916.
FFA 32, Kek Bertincourt, Abwehr Kommando Nord, Jasta 1
9 victories

Leutnant **Herbert Mahn**
Jasta 72
9 victories
In the photo Mahn has the walking stick; on the right is Karl Arnold (6).

Leutnant **Eberhardt Mohnicke**
Jasta 11
9 victories

Leutnant **Hans Karl Müller**
Born Loschwitz, nr. Dresden, 19 July 1892. Died San Antonio, Texas, 23 July 1977.
Kasta 11, Kek Avillers, Jasta 5
9 victories

***Leutnant* Wilhelm Rosenstein**
Born Stuttgart, 28 January 1892. KIFA Rostenberg, South Africa, May 1949.
Jasta 27, Kest 1b, Jasta 40s
9 victories
'Willi' Rosenstein is seated in an Albatros D.III of *Jasta 27*.

***Leutnant* Karl Fritz Schattauer**
Born Tilst, East Prussia, 7 August 1892.
Jasta 23b and 16b
9 victories
Schattauer is posed with his Albatros D.Va of *Jasta 16b*.

***Leutnant* Adolf Schulte**
Born Ludenscheit, 24 December 1894. KIA France, 12 April 1917.
Jasta 12
9 victories

***Leutnant* Georg Weiner**
Jasta 20, Kest 3, Jasta 3
9 victories
Weiner's Fokker D.VII of *Jasta 3* shows his 'W' insignia.

***Vizefeldwebel* Dietrich Averes** (NO PHOTO)
Born Nordhorn, 4 January 1894. Died 21 September 1982.
Jasta Ober-Ost, Jasta 81
10 victories

***Leutnant* Franz Brandt** (NO PHOTO)
Born Minden, 13 February 1893.
Jastas 19, 27 and 26
10 victories

***Leutnant* Herbert Wilhelm Franz Knappe** (NO PHOTO)
Born Glatzischl, 18 January 1891.
Jasta 81
9 victories

***Vizefeldwebel* Karl Pech** (NO PHOTO)
Born Seidau, 9 December 1894. KIA France, 19 May 1918.
Jasta 29
9 victories

***Leutnant zur See* Paul Achilles**
Seefrontstaffel, Marine Feld Jasta V
8 victories
Photo shows Achilles at extreme right, bareheaded. In the center with flying helmet is *Ltn.z.S.* Eberhard Cranz, CO of *MFJ V*, examining his wrecked Fokker D.VII.

***Leutnant* Ludwig Beckmann**
From Westphalia. Died 20 January 1965.
Jastas 6, 48 and 56
8 victories
'Lutz' Beckmann poses with his Fokker D.VII of *Jasta 56*

***Leutnant* Karl Bohny**
Jasta 5, Kest 7, Jasta 17
8 victories

***Leutnant* Aloys *Freiherr* von Brandenstein**
Jasta 49
8 victories

The photo shows pilots of Jasta 49; front row l. to r: *Ltn*. Brandenstein, *Ltn*. Franz Ray (17), *Ltn*. Bolle (*OzbV*), *Ltn*. Josef Jacobs of *Jasta 7* (48), *Ltn*. Hermann Habich (7); Middle row: *Uffz*. Kaendler (behind bush), *Ltn*. Felix Bornträger (?) (2), unknown; Rear: *Ltn*. Kiel, *Vfw*. Duvel, *Flieger* Grotjahn, *Uffz*. Hennies, *Vfw*. Ernst.

Leutnant **Otto Creutzmann**
Born at Spickendorf Borzirk, Halle an der Salle, 8 January 1892. Died in 1943.
Jasta 20, Kest 4b, Jastas 43 and 46
8 victories
In the photo Creutzmann is in the center.

Leutnant **Günther Dobberke**
Jastas 45 and 60
8 victories
Dobberke poses with his personally marked Albatros D.III.

Oberleutnant **Walter Ewers**
Born Lübeck, 11 May 1892. KIA France, 15 May 1918.
FA 26, Jastas 12 and 77b
8 victories

Oberleutnant **Hans-Eberhardt Gandert**
Born Sandberg, 2 September 1892. Died Augsberg, 24 July 1947.
FA 24, Jasta 51, JGr 6
8 victories

***Leutnant* Max Gössner**
Born Weilheim, Bavaria, 25 August 1894.
Kest 31, Jastas 23b and 77b
8 victories
The Albatros D.V belonged to *Ltn*. Heinrich Seywald (6) of *Jasta 23b*.

***Oberleutnant* Adolf Gutkneckt**
Born Badingen, 12 September 1891.
Jasta 43
8 victories
Gutkneckt is in the center, with his ground crew in front of his Albatros-built Fokker D.VII 571/18.

***Leutnant* Johann (Hans) Gottfried von Haebler**
Born Gross Schönau, 17 May 1893. DoW France, 23 March 1918 (mortally wounded 22 March.).
Jasta 36
8 victories

***Offizierstellvertreter* Wilhelm Hippert**
FA 227, Jastas 39 and 74
8 victories

***Unteroffizier* Paul Hüttenrauch**
Jasta 7
8 victories
Picture shows pilots of *Jasta 7* in late 1918 with Jacobs' Fokker Dr.I. L. to r: *Uffz.* Peisker (1), *Uffz.* Paul Hüttenrauch, *Uffz.* August Eigenbrodt (3), *Ltn.* Willi Nebgen (4), *Ltn.* Josef Jacobs (47-48), *Uffz.* Bannenberg, *Ltn.* Wirth, *Ltn.* Rath, unknown.

***Vizefeldwebel* Fritz John Jacobsen**
Born Berlin-Charlottenburg, 7 July 1894. Died Nürnberg, 3 August 1981.
Jastas 17, 9, 31 and 73
8 victories
The aircraft is Albatros D.Va (OAW) 6555/17 of *Jasta 73*.

***Offizierstellvertreter* Wilhelm Kampe**
Born Hofgeisman, 15 March 1888. KIA Belgium, 8 March 1918.
FFA 35, Jastas 8 and 27
8 victories
In this *Jasta 27* photo, 'Willi' Kampe is front left with *Vfw*. Albert Lux (8) on the right. Behind, l. to r: *Vfw*. Max Krauss (4), *Off.Stv*. Paul Leim, and *Ltn*. Ludwig Luer (6).

***Leutnant* Fritz Kieckhafer**
Born Berlin, 15 January 1896. DoW France, 7 June 1918 (mortally wounded 4 May 1918).
Jasta 32b and 29
8 victories
Photo is of *Jasta 32b* summer 1917: front row l. to r: *Ltn*. Wilhelm Eckenberg, *Ltn*. Fritz Kieckhäfer, *Ltn*. Rudolf Windisch (22), *Ltn*. Hans Rolfes (17); Middle: *Vfw*. Kurt Stark (2), *Oblt*. Otto Schmidt (19), *Ltn*. Lothar Zencominierski (3), *Vfw*. Franz Tabaka (1); Rear: *Ltn*. Richler, *Ltn*. Werner Ulmer, *Vfw*. Kurt Petzinna (2), *Vfw*. Konrad Porella and Holsmint.

***Leutnant* Arthur Korff**
Jasta 60
8 victories
This photo shows *Jasta 60* in October 1918. L. to r: *Vfw*. Wunnenberg, *Vfw*. Zimmermann, *Ltn*. Günther Dobberke (8), *Ltn*. Schwager, *Ltn*. Arnd Benzler (9), *Ltn*. Wolff (1), *Ltn*. Karl Ritscherle (8), *Vfw*. Mack, *Vfw*. Bollemann, and *Vfw*. Arthur Korff.

***Oberflugmeister* Karl Meyer**
Born Mulhausen, 29 January 1894. DoI flying accident, Leipzig, 31 December 1917.
See-Flug Station I at Zeebrugge, Marine Feld Jasta I
8 victories

***Leutnant* Kurt Monnington**
Jastas 22,15 and 18
8 victories
Monnington is seated in his camouflaged *Jasta 15* Albatros D.V marked with his skull insignia.

***Leutnant* Otto Parschau**
Pour le Mérite
Born Klutznitz, East Prussia, 11 November 1890. DoW France, 21 July 1916.
FFA 42, FFA 261,Kampfgeschwader 1, FFA32
8 victories
Parschau is ready for flight in this Pfalz *Eindecker*.

***Leutnant* Claus (Fritz Paul) Reimer**
Jasta 26
8 victories
Photo depicts pilots of *Jasta 26* in March 1918. *Ltn.* Riemer is third man standing from the left. For the other names see Fritz Classen (11).

***Leutnant* Wolfram *Freiherr* von Richthofen**
Born Barzdorf, Silesia, 11 October 1895. Died Bad Ischl, Austria, 12 June 1945.
Jasta 11
8 victories

***Leutnant* Karl Waldemar Ritscherle**
Born Karlsruhe, 1 May 1898. KIA England, 24 August 1940 during the Battle of Britain, in an He 111.
Observer Schutzstaffel 8, pilot Jasta 60
8 victories
The photo shows Ritscherle in his battle-scarred OAW-built Fokker D.VII of *Jasta 60*.

***Leutnant* Richard Runge**
Born Hamburg, 22 November 1890. KIA Belgium, 15 November 1917.
Jasta 18
8 victories

***Vizeflugmeister* Karl Scharon**
Marine Feld Jasta II
Died 26 May 1921.
8 victories
In the picture, Scharon stands third from the left. For the names of the other pilots see Alexander Zenses (18).

***Oberleutnant* Hans Schilling**
Born Zackerick, 24 September 1892. KIA France, 4 December 1916.
Observer FFA 5, FFA22
8 victories
The photo shows Schilling standing right, with his pilot Albert Dossenbach.

Leutnant **Viktor Schobinger**
Born Ulm/Donau, 30 April 1893. Died 24 May 1989.
Jasta 12
8 victories

Leutnant **Karl-August von Schönebeck**
Born Bernstadt, Silesia, 19 January 1898. Died 4 September 1989.
Jastas 11, 59 and 33
8 victories

***Leutnant* Fritz Thiede**
Born Lichterfelde, 21 January 1896.
Jastas 24s and 38
8 victories

***Leutnant* Gerhard Bassenge**
Born Ettlingen, 18 November 1897. Died Lubeck, 13 March 1977.
Jasta Boelcke
7 victories
Bassenge sits atop his Albatros D.III of *Jasta Boelcke*.

***Oberleutnant* Theodor Hermann Dahlmann**
Born Wanfried, Hessen, 19 November 1892. Died 21 January 1978.
Jastas 29 and 26, JG Nr.III
7 victories
Dahlmann is on the left with Hermann Göring.

***Leutnant* Helmut Dilthey**
Born Rheydt, nr. Düsseldorf, 9 February 1894. KIA France, 9 July 1918.
Jastas 27 and 40
7 victories
Dilthey is seated in his *Jasta 40* Albatros D.Va.

***Leutnant* Julius Fichter**
Born Billinghain, 9 November 1891. DcI Graudenz, 13 December 1918 (Badly injured in an aeroplane accident 9 December 1917.).
Kasta 26, Jastas 22 and 67
7 victories
In this photo of *Jasta 22*, *Ltn*. Fichter is 4th from the left. For the other names see Bohenkamp (15).

***Vizeflugmeister* Hans Goerth**
Marine Feld Jastas I and III
7 victories
The aircraft is a Fokker E.V of *MFJ III*.

Leutnant **Gisbert-Wilhelm Groos**
Born 10 August 1894.
Jastas 4 and 11
7 victories

Leutnant **Hermann Habich**
Born Plattig, nr. Buhl, 15 August 1895.
Jasta 49
7 victories

Habich's checkered *Jasta 49* machine is Albatros D.III (OAW) 5127/17, although the serial number on the wheel shows it came from 5129/17.

***Hauptmann* Otto Hartmann**
Born Nassau, Württemberg, 7 February 1889. KIA Belgium, 3 September 1917.
Observer Kampfstaffel 15, pilot Jastas 18 and 28w
7 victories

***Leutnant* Georg *Ritter* von Hengl**
Born October 1897. Died 19 March 1952.
FA 295
7 victories
Picture shows *Ltn.* Hengl, left, with his gunner, Johann Baur (6) by their Hannover CL.IIIa.

Vizefeldwebel **Josef Hohly**
Jastas 29 and 65
7 victories

The picture shows a line-up of *Jasta 65* pilots in May 1918. From l. to r: *Ltn.* Otto Fitzner (9), *Ltn.* Wilhelm Frickart (12), *Ltn.* Heinrich Zempel (1), *Ltn.* Bieber, *Ltn.* Heinz *Freiherr* von Beaulieu-Marconnay, *Ltn.* Walter Fritzsche, *Off. Stv.* Tiedje (3), *Vfw.* Josef Hohly, *Gfr.* Scheutzel, *Ltn.* Rost (*OzbV*).

Leutnant **Martin Johns**
Jasta 63
7 victories

Johns is seen with his Albatros D.Va marked with the black diamond pattern of *Jasta 63*, and his personal 'J' insignia.

Oberflugmaat **Christian Kairies**
Born Memel, 24 October 1896. DoW Zeebrugge, 2 October 1918 (mortally wounded 1 October.).
Seefrontstaffel, Marine Feld Jasta V
7 victories

Leutnant **Emil Koch**
Jastas 12 and 32b
7 victories

***Leutnant* Hans Körner**
FA 246, Jastas 8 and 19
7 victories
Körner is seen with his *Jasta 19* Fokker Triplane,Dr.I 503/17.

***Offizierstellvertreter* Wilhelm Kühne**
Born 11 December 1888. KIA 30 August 1918.
Jastas 29, 15 and 18
7 victories
Kühne is seen fourth from left, with ground personnel of *Jasta 18* at Montingen in the summer of 1918. The Fokker D.VII bore the red and white colors and 'raven' emblem of this famous '*Rabenstaffel*'.

***Oberleutnant* Hans Kummetz**
Born Illowo, 29 April 1890. KIA Italy, 11 January 1918.
Jasta 1
7 victories
Kummetz is in the center of this *Jasta 1* group, with the cigar.

***Leutnant* Hermann Leptien**
Jasta 21s and 63
7 victories
Pilots from *Jasta 63* pose in front of an Albatros D.Va displaying the unit's black diamond markings. L. to r: *Uffz*. Jon Santjer (2), *Ltn*. Martin Johns (7), *Ltn*. Hermann Leptien, unknown, *Uffz*. Ruppert Merkle, *Uffz*. Friedrich Neubauer (4), unknown.

***Vizefeldwebel* Albert Lux**
Jasta 27
7 victories

***Leutnant* Karl Albert Mendel**
Born Jerichow, 4 October 1892. KIA France, 6 June 1918.
Jastas 15 and 18
7 victories

***Leutnant* Alfred Niederhoff**
Born Dortmund, 2 January 1895. KIA Belgium, 28 July 1917.
Jastas 20 and 11
7 victories
This picture of *Jasta 20* shows *Ltn*. Alfred Neiderhoff (top left), *Ltn*. Gerhard Flecken (4) and *Ltn*. Friedrich Mallinckrodt (6), while *Oblt*. Fritz Heissing the CO is in the center with hands clasped.

***Hauptmann* Paul Henning Aldabert Theodor von Osterroht**
Born Luneberg, 13 September 1887. KIA France, 23 April 1917.
Jasta 12
7 victories
Von Osterroht stands at right with his former observer (in BAO) and now famous fighter pilot Manfred von Richthofen.

***Leutnant* Richard Plange**
Born Ellingson, 24 December 1892. KIA Belgium, 19 May 1918.
Jastas Boelcke and 36
7 victories

The picture is of pilots of *Jasta* (2) *Boelcke* in early 1918. L. to r: *Oblt.* R. von Griesheim (1), *Ltn.* Otto Löffler (15), *Ltn.* Karl Gallwitz (9), *Ltn.* Hermann Vallendor (6), *Oblt.* Karl Bolle (36), *Ltn.* Fritz Kempf (4), *Ltn.* Harry von Bülow-Bothkamp (6), *Ltn.* Richard Plange, *Ltn.* Paul Schröder (1) and *Vfw.* Paul Bäumer (43).

***Vizefeldwebel* Johann Pütz**
Jastas 23b and 34b
7 victories

***Leutnant* Josef Raesch**
Born Zewen, nr. Trier, 4 June 1897.
Jasta 43
7 victories
Rasech's victory score is disputed, perhaps only 3 or 4 actually confirmed by *Kogenluft*. He stands with his *Jasta 43* Fokker D.VII.

***Vizefeldwebel* Gustav Schneidewind**
Born 28 August 1890.
Jastas 17 and 1F in Palestine
7 victories

Vizefeldwebel **Georg Strasser**
Jasta 17
7 victories

Unteroffizier **Karl Treiber**
Born in 1895.
Jasta 5
7 victories
Treiber is seen with his Fokker D.VII, which was reportedly marked with a 'T'.

***Hauptmann* Franz Josef Walz**
Pour le Mérite
Born Speyer, nr Mannheim, 4 December 1885. Died Breslau, December 1945 (as a Russian PoW).
Kampfstaffel 2, Jasta Boelcke and 34b, FA 304b
7 victories

***Leutnant* Johannes (Hans) Werner**
Jasta 14
7 victories

Leutnant **Konrad Brendle** (NO PHOTO)
Born Vollfosen, 10 August 1895. KIA France, 2 September 1918.
Jastas 1 and 45
8 victories

Vizefeldwebel **Friedrich Ehmann** (NO PHOTO)
Kest 5, Jasta 47w
8 victories

Leutnant **Wolfgang Güttler** (NO PHOTO)
Born Reichenstein, 22 February 1893. KIA France, 20 February 1918 (collision during an air battle.).
Jasta 24s and 13
8 victories

Leutnant **Heinrich Henkel** (NO PHOTO)
Born Hamburg, 11 May 1896.
Jasta 37
8 victories

Leutnant **Hans Hoyer** (NO PHOTO)
Born Rostock, 20 September 1890. KIA France, 15 November 1917.
Jasta 36
8 victories

Unteroffizier **Michael Hutterer** (NO PHOTO)
Born Munich, 17 November 1891.
Jasta 23b
8 victories

Leutnant **Franz Kirchfeld** (NO PHOTO)
Jasta 73
8 victories

Vizefeldwebel **Fritz Krebs** (NO PHOTO)
Born Cottbus, 26 November 1896. KIA Belgium, 16 July 1917.
Jasta 6
8 victories

Vizefeldwebel **Friedrich Poeschke** (NO PHOTO)
Jasta 53
8 victories

Leutnant **Wilhelm Schwartz** (NO PHOTO)
Jastas 20, 69, 13 and 73
8 victories

Unteroffizier **Wilhelm Zorn** (NO PHOTO)
Jasta 60
8 victories

Leutnant **Friedrich Gerhard Anders** (NO PHOTO)
Jastas 4 and 73
7 victories

Leutnant **Helmut Brünig** (NO PHOTO)
Jastas 32b, 50 and 41
7 victories

Leutnant **Kurt Jacob** (NO PHOTO)
Born Schleiz, 7 June 1896. KIFA Hannover, 10 March 1919.
Jastas 33 and 36
7 victories

Unteroffizier **Marat Schumm** (NO PHOTO)
Schutzstaffel 26, Jasta 52
7 victories

Vizefeldwebel **Kurt Ungewitter** (NO PHOTO)
Born Elbersfeld, 13 November 1891. KIFA 14 March 1927.
Schusta 5 and Jasta 24
7 victories

***Vizefeldwebel* Karl Arnold**
Jasta 72s
6 victories
Karl Arnold stands on the right of this group, with Herbert Mahn (9) holding the walking stick.

***Unteroffizier* Johann Baur**
Born Muhldorf, 24 April 1897. Died in 1993.
Observer FA 295
6 victories

***Leutnant* Harry von Bülow-Bothkamp**
Born Bothkamp Estate, Holstein, 19 November 1897. Died Bothkamp Estate, 27 February 1976.
Jastas 36, Boelcke, and 36
6 victories

***Oberflugmeister* Albin Buhl**
Seefrontstaffel, Marine Feld Jasta IV
6 victories

Leutnant August Burkard
Jasta 29
6 victories
This *Jasta 29* photo was taken in late 1918. 5th from the left is *Oblt*. Harold Auffarth (29) commander of *Jasta 29*, while *Ltn*. August Burkard is 7th from the left. The aircraft are Pfalz D.VIII fighters.

Leutnant Alfred Fleischer
From Silesia. Died Germany, 11 June 1978.
Jasta 17
6 victories
Fleischer is seen in the cockpit of his Albatros-built Fokker D.VII of *Jasta 17*.

***Leutnant* Gustav Frädrich**
FA 30, Jastas 39, 1 and 72s
6 victories
Frädrich is assisted with his 'Heinecke' parachute harness and pack, while his Fokker D.VII 383/18 awaits in the background, at Bergnicourt in July 1918.

***Leutnant* Sylvester Garsztka**
Born Bydgaszey (in what is now Poland), 31 December 1896. KIFA 8 June 1919.
Jasta 31
6 victories

Unteroffizier Friedrich Gille
Jasta 12
6 victories

Jasta 12 pilots at their ease: standing, l. to r: *Vfw*. Otto Rosenfeld (13), *Ltn*. Karl Schöck (2), *Ltn*. Friedrich Hochstetter (1), *Uffz*. Friedrich Gille, *Ltn*. Oskar Müller (*OzbV*), *Vfw*. Robert Riessinger (4), *Vfw*. Reinhold Jörke (13); Seated: *Oblt*. Adolf von Tutschek (27), *Ltn*. Bender, *Vfw*. Grigo, *Vfw*. Arthur Schorisch (4) and *Ltn*. Paul Billik (31).

Leutnant Hermann Gilly
Born Donaueschingen, 16 September 1894.
Jasta 40s, 29 and 40s
6 victories

Hermann Gilly appears bareheaded in the middle of this group of *Jasta 40* pilots in late 1918. Seated, l. to r.: *Ltn*. Frodien, *Ltn*. Carl Degelow (30), unknown, *Ltn*. Hermann Gilly, *Ltn*. Willy Rosenstein (9), *Ltn*. Anton Raab, and *Ltn*. Hans Jeschonnek.

***Unteroffizier* Heinrich Haase**
Schlasta 8, Jasta 21s
6 victories

Picture of *Jasta 21s* at the end of August 1918. Front l. to r: *Oblt.* von Kalkreuth (*OzbV*), *Oblt.* Oskar *Fr.* von Boenigk (26), *Ltn.* Fritz Höhn (20); Middle row: *Vfw.* Karl Schmückle (6), *Ltn.* Christians, *Ltn.* Julius Keller (1), unknown; Third row: two unknowns; Last row: *Uffz.* Heinrich Haase, *Ltn.* Schlumps, *Vfw.* Karl Thom (27).

***Oberleutnant* Erich Hahn**
Born Leipzig, 18 October 1891. KIA France, 4 September 1917.
FFA 23, FA (A) 221, Kek Bertincourt, Jastas 1 and 19
6 victories

***Leutnant* Otto Walter Höhne**
Born Wiowitz, nr. Ratibor, 30 April 1895. Died 22 November 1969.
Jastas 1, 2 and 59
6 victories

***Vizefeldwebel* Alfred Hübner**
Born Lauenberg, 26 October 1891.
Jasta 36
6 victories

This early 1918 photo of *Jasta 36* depicts in the front row l. to r: *Vfw*. Meyer, *Ltn*. Wandelt (?), *Ltn*. Heinrich Bongartz (33), *Ltn*. Max Naujock, *Ltn*. Max Fuhrmann and *Vfw*. Alfred Hübner. In the second row, *Vfw*. Patzer is on the wing left of the cowling of the Fokker Dr.I with *Ltn*. Hans von Haebler (8) on the right.

***Leutnant* Hans Imelmann**
Born Hannover, 14 May 1897. KIA France, 23 January 1917.
Kek Metz, Jasta 2
6 victories

***Leutnant* Johannes Jensen**
Jasta 57
6 victories
Jasta 57 pilots line up in front of their Fokker D.VIIs. From l. to r: *Uffz.* Tracinski (2), *Vfw.* Hanzog (1), *Gfr.* Hechler (2), *Ltn.* Johannes Jensen, *Ltn.* Paul Strähle (14), *Hptm.* Seldner, *Ltn.* Blum (3), *Vfw.* Wieprich (4), *Off. Stv.* Dudel, *Uffz.* Hitschler.

***Leutnant* Erich Just**
Born in 1898.
Jasta 11
6 victories

***Feldwebel* Wilhelm Kahle**
Born Tarff in Sans, 13 April 1892.
Jasta 27
6 victories

***Vizefeldwebel* Max Kahlow**
Born Bad Schonfliess, 11 May 1894.
FA42, Jasta 34b
6 victories
Kahlow is seen in a *Jasta 34b* Albatros D.V.

***Vizefeldwebel* Gustav Klaudat**
From East Prussia.
Jasta 15
6 victories
This photo shows pilots of *Jasta 15* at Charmois, near Stenay, celebrating Georg von Hantelmann's 20th victory on 9 October 1918. L. to r: *Vfw*. Gustav Klaudat, *Ltn* Glöcklen, *Ltn*. Joachim von Ziegesar (3), *Ltn*. Josef Veltjens (35), *Ltn*. Georg von Hantelmann (26), *Ltn*. Hugo Schäfer (11), and *Vfw*. Theodor Weischer (4).

***Leutnant* Erich König**
Born Oberhausen, 2 April 1890. KIA France, 2 April 1917.
Jasta 2
6 victories

Pilots of *Jasta Boelcke*, late 1916. L. to r: *Off. Stv*. Max Müller (36), *Ltn*. Manfred *Frhr*. von Richthofen (80), *Ltn*. Wolfgang Günther, *Oblt*. Stefan Kirmaier (11), *Ltn*. Hans Imelmann (6), *Ltn*. Erich König, *Ltn*. Otto Höhne (6), *Ltn*. Hans Wortmann (2), and *Ltn*. Dieter Collin (13).

***Leutnant* Hermann Kunz**
Jastas 3, 7 and 1F in Palestine
6 victories

Kunz is seen seated in his black *Jasta 7* Albatros D.V with his personal swastika marking.

***Leutnant* Kurt Küppers**
Born 2 December 1894. Died 24 June 1971.
Jasta 6, Bogohl 3, Jasta 48
6 victories
In between *Jastas*, Küppers flew Gotha G.IV bombers in *Staffel 16* of *Bogohl 3*, and is seen here 2nd from the left with *Oblt*. Fritz Lorenz third from left.

***Leutnant* Alfred Lenz**
Born Munich, 15 March 1890. Died Hannover, 10 May 1965.
FA 61, Jastas 4, 14 and 22
6 victories
The aircraft is Lenz' favorite Siemens Schuckert D.IV, 3083/18.

Leutnant **Ludwig Luer** (rear)
Jastas 27 and 62
6 victories
Luer is standing directly behind *Vfw*. Albert Lux (8) of *Jasta 27*.

Leutnant **Friedrich Mallinckrodt**
Born Essen, 15 August 1894. Killed during WW2.
Kek Sivry, Jastas 6 and 20
6 victories
Mallinkrodt is seen with an Albatros D.II.

Leutnant **Heinrich Maushake**
Born 7 December 1894. Died 31 August 1981.
Jasta 4
6 victories
Maushake examines a pocket camera, with a *Jasta 4* Fokker D.VII in the distance at Metz-Frescaty airfield.

Leutnant **Emil Meinecke**
Born Mannheim, 20 July 1892. Died Canada, 2 May 1975.
Ottoman FA 6
6 victories

***Leutnant* Werner Niethammer**
Died in 1946.
Jasta 13
6 victories

***Leutnant* Arthur Rahn**
Born Schweingrube, East Prussia, 18 July 1897.
Jastas 19, 15, and 19
6 victories
Rahn is seated in his *Jasta 19* Fokker Dr.I 433/17.

***Leutnant* Rudolf Rienau**
Born Sylt Island, 17 November 1898. KIFA Berlin, 23 May 1925.
Jasta 19
6 victories

***Feldwebel* Karl Schmückle**
Jastas 21s and 15
6 victories
In this *Jasta 21* photo, *Fw*. Karl Schmückle is in the center. *Ltn*. Julius Keller (2) is thought to be on the left, and *Vfw*. Karl Thom (27) is on the right beside the spinner of the Albatros D.V.

***Leutnant* Edgar Scholtz**
Born Suhl, Thuringia, 21 September 1898. KIFA Cappy, France, 2 May 1918.
Kampfstaffel 10, Jasta 11
6 victories

***Leutnant* Wilhelm Schulz**
Born Strasburg, 22 May 1897. KIA France, 25 June 1918.
Jastas 16b, 41 and 66
6 victories
The photo is of pilots of Bavarian *Jasta 16*. Standing l. to r: *Ltn*. Ferdinand Gröner, *Ltn*. Ludwig Hanstein (16), *Ltn*. Eberhard Baier, 'Willi Schulz, *Ltn*. Theodor Rumpel (5); Seated: *Ltn*. Karl Odebrett (14), *Ltn*. Bruno Breitenbach, and *Hptm*. Paul Kremer (2).

***Leutnant* Günther Schuster**
Jastas 17, 29 and 17
6 victories
Schüster is seen with the Albatros-built Fokker D.VII he flew as commander of *Jasta 17*.

***Offizierstellvertreter* Erich Schütze**
FA 201, Rhz9, Jasta 25
6 victories

Leutnant **Heinrich Seywald**
Born Regensburg, 23 December 1894.
Jasta 23b
6 victories
Seywald is standing in front of his personal Albatros D.V marked with a black 'S' on a white-bordered black fuselage band. The marking of *Jasta 23b* is seen on the tail.

Vizefeldwebel **Erich Sonneck**
Jasta 66
6 victories
In the background is the Fokker D.VII (OAW) 2035/18 of Rudolf Winc[illegible]sch (22) of *Jasta 66*, marked with his white stag and yellow shield emblem.

***Oberleutnant* Kurt Student**
Born Birkholz, Brandenburg, 12 May 1890. Died Germany, 1 July 1978.
FA17, AOK3, Jasta 9
6 victories

***Leutnant* Hermann Stutz**
Died 27 July 1938.
Jastas 20 and 71
6 victories
Stutz center, with Friedrich Mallinckrodt (6) behind on the left.

Offizierstellvertreter **Reinhard Treptow**
Born Leikow, Krikolberg, 28 November 1892.
Jasta 25
6 victories

Leutnant **Hermann Vallendor**
Born Offenberg, 13 April 1894.
Jastas 26 and Boelcke
6 victories

***Oberleutnant* Hans Waldhausen**
Born Mainz/Rhein, 30 January 1892. Died 6 November 1976.
Jasta 37
6 victories

***Vizefeldwebel* Ernst Wiehle**
Born Harzgerode, 24 November 1894.
Schutzstaffel 3, Jasta 43
6 victories
Wiehle is seated in his *Jasta 43* Fokker D.VII.

***Leutnant* Hans Auer**
Bavarian
Jastas 26 and 32b
5 victories

***Leutnant* Joachim von Bertrab**
Born St. Andreasberg, 18 February 1894. Died Boizenburg, 28 July 1922.
Jasta 30
5 victories
Bertrab is posed in front of his black Albatros D.III of *Jasta 30*.

***Oberleutnant* Hans-Helmut von Boddien**
Born Altenplatow, Saxony, 26 October 1890. KOAS Baltic, 29 November 1919.
Jastas 11 and 59
5 victories

***Leutnant zur See* Erich Bönisch**
Observer Seaplane Station 1, Flanders
5 victories

***Vizefeldwebel* Hans Bowski**
Jastas 14 and 51
5 victories

Jasta 14 pilots gather in the Officers' Kasino; Front l. to r: *Ltn.* Kuen (2), *Oblt.* Rudolf Berthold (44), *Vfw.* Hans Bowski, *Off. Stv.* Hüttner (3), *Vfw.* Otto Gerbig (4), *Ltn.* Walter Kypke (9), *Ltn.* P. Lohmann; Back row: *Ltn.* Josef Veltjens (37), *Ltn.* Breuer (1), *Ltn.* Friedrich Vonschott, *Vfw.* Hermann Margot, unknown, *Ltn.* Alfred Lenz (6).

***Leutnant* Herbert Boy**
Jasta 14
5 victories

Picture taken at Boncourt airfield in March 1918 with *Jasta 14* Fokker Dr.I – and mascot!

***Leutnant* Otto von Breiten-Landenberg**
Jastas 9 and 6
5 victories
This photo commemorated a visit to *Jasta 9* by members of the Dresden Opera Company. *Ltn.* Otto Breiten-Landenberg stands third from the right, and *Ltn.* Herbert Rolle (3) is third from the left.

***Hauptmann* Friedrich-Karl Burckhardt**
Born Koslin, 24 December 1889. Died 18 June 1962.
Jasta 25
5 victories
Burkhardt was photographed in Macedonia, with a *Jasta 25* Albatros D.III.

***Leutnant* Karl Christ**
Born Darmstadt, 15 June 1897.
Jasta 28w
5 victories

Pilots with *Jastas 28w* and *33* (part of *JGr.7*) gather in the summer of 1918. Seated front, l. to r: *Ltn.* Otto Koch (*Jasta 28 OzbV*), *Ltn.* Sagebühl, *Ltn.* Karl Christ; Rear: *Ltn.* Karl August von Schönebeck (CO *Jasta 33*, with 8 victories), *Ltn.* Schröder (*Jasta 28*), *Ltn.* Emil Thuy (CO *Jasta 28*, with 32 victories), *Ltn.* Busse (*Jasta 28*), *Ltn.* Rebbe, and *Ltn.* Richard Ruckle (2) of *Jasta 33*.

***Oberleutnant* Theodor Jakob Croneiss**
Died 7 November 1942.
Ottoman FFA6
5 victories

***Offizierstellvrtreter* Wilhelm Cymera**
Born Elberfeld, Westphalia, 5 October 1885. KIA France, 9 May 1917.
Jasta 1
5 victories

***Leutnant* August Delling**
Born Braunetsried, Bavaria, 19 October 1895. Died 17 June 1962.
Jasta 34b
5 victories

***Leutnant* Wilhelm Fahlbusch**
Born Hannover, 8 February 1892. KIA France, 6 September 1916.
Kasta 1
5 victories

***Leutnant* Siegfried Gussmann**
Born in 1896.
FA 3, Jasta 11
5 victories

***Leutnant* August Hanko**
Jastas 28w and 64w
5 victories
Hanko is seated in a *Jasta 28w* Albatros D.III, which was photographed with many other pilots.

***Leutnant* Kurt Hetze**
Jasta 13
5 victories

***Vizefeldwebel* Karl Friedrich Kurt Jentsch**
FA 66, FA 30, Jasta 1, FA (A) 234, Jastas 61 and Boelcke
5 victories
Jentsch is in the cockpit of his *Jasta 61* Albatros D.Va 2343/17, emblazoned with his personal star emblem.

***Vizefeldwebel* Hermann Juhnke**
Born Laurenberg, 7 April 1893.
Jastas 41 and 52
5 victories

***Leutnant* Werner Junck**
Born Magdeburg, 28 December 1895. Died Münich, 6 August 1976.
Jasta 8
5 victories
This photograph, the only one available to the authors, was obviously taken during World War II.

***Leutnant* Wilhelm Kohlbach**
Jastas 50 and 10
5 victories
Photo of *Jasta 10* in front of their Fokker D.VIIs, at immediate readiness: l. to r: *Uffz.* Oskar Henning, *Ltn.* Kurt Schibilsky (1), *Ltn.* Justus Grassmann (10), *Ltn.* Aloys Heldmann (15), *Off. Stv.* Paul Aue (10), *Ltn.* 'Willy' Kohlbach, *Uffz.* Rudolf Klamt, *Ltn.* Gustav Böhren.

***Leutnant* Wilhelm Leusch**
Born Neuss, nr. Dusseldorf, 15 October 1892. KIFA 14 August 1921.
Jastas 13 and 19
5 victories
Leusch is seated in an Albatros D.II, marked with his 'L' in typical *Jasta 19* style.

***Oberleutnant* Heinrich Lorenz**
Born in Saxony, 30 January 1892.
KGI, Jastas 1 and 33
5 victories
Picture shows Lorenz as CO of *Jasta 33*, at left rear. The pilot at front on the right is *Ltn*. Friedrich Ermann (3 victories).

***Unteroffizier* Hans Heinrich Marwede**
Jasta 67
5 victories
Marwede is seen just after his capture by U.S. troops on 3 October 1918.

***Leutnant* Hans-Georg von der Osten**
Born Rittergut Witzmitz, East Pomerania, 9 September 1895. Died 17 March 1987.
Jasta 11 and 4
5 victories

Offizierstellvertreter **Leopold Rudolf Reimann**
Born Oberhohnsdorf, near Zwickau, Saxony, 7 December 1892. KIFA France, 24 January 1917.
Jastas 1 and 2
5 victories

Leutnant **Hans Rosencrantz**
Born Wollstein, 9 August 1890. KIA France, 6 September 1916.
Observer KG 1
5 victories
His pilot was *Ltn*. Willi Fahlbusch (5).

Vizefeldwebel **Paul Rothe**
Born 20 July 1890.
Jasta 14
5 victories

Leutnant **Theodor Rumpel**
Born Bahrenfeld, nr. Hamburg, 25 March 1897.
Jastas 16b and 23b
5 victories
Rumpel is seen with his *Jasta 23b* Albatros D.V, which displayed the *Staffel* marking on the tail and his personal green finish on the fuselage.

***Leutnant* Herbert Schröder**
Jastas 1 and 17
5 victories
Schröder stands at the back; *Oblt.* Herbert Theurich, CO *Jasta 35b* is in front.

***Vizefeldwebel* Friedrich Schumacher**
Jasta 10
5 victories
Friedrich Schumacher may be the pilot second from right in this *Jasta 10* photo, but some historians say this was actually *Vfw.* Burggaller. Known identifications are, l. to r: *Ltn.* Justus Grassman (10), *Ltn.* Aloys Heldmann (15), *Oblt.* Hugo Schäfer (*OzbV*), *Ltn.* Erich Löwenhardt (CO, with 54), *Ltn.* Fritz Friedrichs (21), *Ltn.* Walter Lehmann, unknown, Paul Aue (10), Schumacher (?), *Vfw.* Kurt Delang (3).

***Leutnant* Kurt Seit**
Born 30 July 1894
Jasta 80
5 victories
Kurt Seit is seen in his highly decorated Fokker D.VI of *Jasta 80b*, named 'Fratz'.

***Vizefeldwebel* Wilhelm Stör**
Born Witten, Ruhr, 5 October 1893. Died Augsberg, 12 December 1977.
Jasta 68
5 victories

Vizefeldwebel **Karl Strünkelnberg**
Jasta 9
5 victories

***Vizeflugmeister der Marineinfanterie* Wilhelm Thöne**
Seeflieger Abteilung II, Marine Feld Jasta I
5 victories
There is some dispute over this score; Thöne definitely had four confirmed victories and at least one 'forced to land', and more unconfirmed.

Leutnant **Alwin Thurm**
Born Hagan, Westphalia, 11 March 1894. KIA Italy, 31 December 1917.
Jastas 24s and 31
5 victories
The aircraft is an Albatros D.V of *Jasta 31*.

Leutnant **Gerold Tschentschel**
FA 37, Jasta 72s
5 victories
Tschentschel is on the right, Gustav Frädrich (6) is to the left with walking stick.

***Leutnant* Alfred Ulmer**
Born Reutlingen, nr. Stuttgart, 15 September 1896. KIA France, 29 June 1917.
Jasta 8
5 victories
Ulmer is seen with a captured Sopwith Pup.

***Oberleutnant* Hasso von Wedel**
Born Vogelsang, 12 May 1893. KOAS Berlin, 1 May 1945.
Observer KG 5, pilot Jastas 14, 75 and 24s.
5 victories
The Fokker D.VII with von Wedel's family coat of arms of an East Prussian 'Richtrad; was probably flown in *Jasta 24s*.

***Leutnant* Ernst Wiessner**
Born Stuttgart, 1 December 1894. KIA Belgium, 7 June 1917.
FFA 250, Jasta 18
5 victories

***Leutnant* Kurt Wissemann**
Born Elberfeld, 20 March 1893. KIA France, 28 September 1917.
Jasta 3
5 victories

***Hauptmann* Martin Zander**
Died 1925.
FFA 90, Kek Nord, Jasta 1
5 victories

***Leutnant* Paul Bona** (NO PHOTO)
Born Berlin, 1 January 1895. KIA France, 6 June 1917.
Jasta 1
6 victories

***Leutnant* Moritz-Waldemar Bretschneider-Bodemer** (NO PHOTO)
Born Harthau, 29 May 1892. KIA France, 18 July 1918.
Jasta 6
6 victories

***Leutnant* Karl Deilmann** (NO PHOTO)
Born 22 April 1894.
Jasta 6
6 victories

***Flugmaat* Karl Engelfried** (NO PHOTO)
From Saxony, born c. 1893.
Seefrontstaffel, Marine Feld Jasta V
6 victories

***Leutnant* Albert Hets** (NO PHOTO)
Born Hamburg, 5 March 1896. KIA France, 10 August 1918.
Jasta 37
6 victories

***Offizierstellvertreter* Robert Hildebrandt** (NO PHOTO)
Born 19 June 1893.
Jastas 13, 12, 69 and 53
6 victories

***Vizefeldwebel* Otto Klaiber** (NO PHOTO)
FA 232, Jasta 12
6 victories

***Oberleutnant zur See* Konrad Mettlich** (NO PHOTO)
Born Münster, 16 January 1894. KIA France, 13 March 1918.
Marine Feld Jasta I, Jasta 8
6 victories

***Leutnant* Alfred Mohr** (NO PHOTO)
Born Oberremmel, 20 January 1889. KIA France, 1 April 1917.
Jasta 3
6 victories

***Leutnant* Hans Oberländer** (NO PHOTO)
FA 71, Jasta 30
6 victories

***Leutnant* Rudolf Otto** (NO PHOTO)
FA 283, Jastas 74 and 68
6 victories

***Leutnant* Alfons Scheicher** (NO PHOTO)
Schutzstaffel 27, Jasta 34b
6 victories

***Leutnant* Otto Splitgerber** (NO PHOTO)
Born Schwaben, 26 August 1883. KIA Macedonia, 13 March 1918.
Jasta 12 and 38
6 victories

***Vizefeldwebel* Georg Staudacher** (NO PHOTO)
Born 1 August 1898. Died 15 October 1981.
Jasta 1
6 victories

***Unteroffizier* Sigfried Westphal** (NO PHOTO)
Jasta 29
6 victories

***Unteroffizier* Bernard Bartels** (NO PHOTO)
Jasta 44s
5 victories

***Unteroffizier* Rudolf Besel** (NO PHOTO)
Born Bronnen, 26 May 1895. KIA France, 7 September 1918.
Schutzstaffel 30b
5 victories

***Vizefeldwebel* Otto Bieleit** (NO PHOTO)
Jastas 45 and 66
5 victories

***Flugmaat* Eduard Blaas** (NO PHOTO)
Marine Feld Jastas II and III
5 victories

***Unteroffizier* Gustav Borm** (NO PHOTO)
Jasta 1
5 victories

***Vizefeldwebel* Ludwig Gaim** (NO PHOTO)
Born Daggendorf, 1 April 1892.
Jasta 39
5 victories

***Leutnant* Johannes Gildemeister** (NO PHOTO)
Jasta 20
5 victories

***Leutnant* Kurt Haber** (NO PHOTO)
From Beithen, Upper Silesia, born 4 July 1895. KIA France, 20 December 1916.
Jastas 15 and 3
5 victories

***Leutnant* Friedrich Hengst** (NO PHOTO)
Schusta 3, Jasta 64w
5 victories

***Offizierstellvertreter* Johann Kopka** (NO PHOTO)
FA262, Jasta 61
5 victories

***Unteroffizier* Erich Meyer** (NO PHOTO)
Born Elberstadt, 15 March 1898.
Jasta 45
5 victories

***Vizefeldwebel* Richard Rübe** (NO PHOTO)
Jasta 67
5 victories

***Gefreiter* Johann Schlimpen** (NO PHOTO)
Born Daun, nr. Cologne, 11 March 1895. KIA France, 20 August 1918.
Jasta 45
5 victories

***Leutnant* Roman Schneider** (NO PHOTO)
Born Passau, Bavaria, 9 August 1898.
Jasta 79
5 victories

***Leutnant* Konrad Schwartz** (NO PHOTO)
Died as PoW, 24 August 1918 (shot down, wounded, and taken prisoner 18 July 1918.)
Jastas 22s and 66
5 victories

***Leutnant* Eugen Siempelkamp** (NO PHOTO)
Jastas 4, 29 and 64w
5 victories

***Leutnant* Wilhelm Sommer** (NO PHOTO)
Jasta 39
5 victories

***Leutnant* Helmut Steinbrecher** (NO PHOTO)
Born Dresden, 30 June 1896.
Jasta 46
5 victories

***Vizefeldwebel* Oswald Tränker** (NO PHOTO)
Observer Schusta 16
5 victories

***Leutnant* Hans Viebig** (NO PHOTO)
Jastas 20 and 57
5 victories

***Leutnant* Werner Wagener** (NO PHOTO)
Born Dessau, 13 November 1894.
Jastas 21s and 39
5 victories

Name	*Rank*	*Main units*	*Score*	*Page*
Achilles, Paul	*Ltn.z.S.*	*SFS, MFJ V*	8	108
Adam, Hans von	*Ltn.*	*J34, 6*	21	142
Allmenröder, Karl	*Ltn.*	*J11*	30	26
Altemeier, Friedrich	*Vfw.*	*J14, 24*	21	43
Althaus, Ernst *Fr* von	*Rittm.*	*J4, 14, 10*	9	99
Anders, Fritz Gerhard	*Ltn.*	*S8, J35, 4, 73*	7	137
Anslinger, Leopold	*Ltn.*	*FA54, 24*	10	91
Arnold, Karl	*Vfw.*	*J72*	6	138
Arntzen, Heinrich	*Ltn.*	*J15, 20, 50*	11	83
Aue, Paul	*Off.Stv.*	*J10*	10	91
Auer, Hans	*Ltn.*	*J16, 26, 32*	5	160
Auffarth, Harald	*Oblt.*	*J18, 29*	29	29
Averes, Dietrich	*Vfw.*	*J81*	10	107
Baldamus, Hartmut	*Ltn.*	*J5, 9*	18	52
Barnekow, Raven *Fr* von	*Ltn.*	*J20, 1*	11	84
Bartels, Bernard	*Uffz.*	*J44*	5	181
Bassenge, Gerhard	*Ltn.*	*J5, 2*	7	122
Bäumer, Paul	*Ltn.*	*J2, 5, 2*	43	16
Baur, Johann	*Uffz.*	*FA295*	6	138
Beaulieu-Marconnay, Oliver *Fr* v	*Ltn.*	*J18, 15, 19*	25	36
Becker, Hermann	*Ltn.*	*J12*	23	39
Beckman, Ludwig	*Ltn.*	*J6, 48, 56*	8	108
Benzler, Arno	*Ltn.*	*J32, 45, 65, 60*	9	100
Bernert, Otto	*Oblt.*	*J4, 2, 6*	27	32
Berr, Hans	*Oblt.*	*Kek A, J5*	10	92
Berthold, Rudolf	*Hptm.*	*Kek V, J4, 18 JGII*	44	15
Bertrab, Joachim von	*Ltn.*	*J30*	5	180
Besel, Rudolf	*Uffz.*	*S30*	5	181
Bethge, Hans	*Oblt.*	*J1, 30*	20	45
Bieleit, Otto	*Vfw.*	*J45, 60*	5	181
Billik, Paul	*Ltn.*	*J12, 7, 52*	31	25
Blaas, Eduard	*Flgm.*	*MFJII, III*	5	181
Blume, Walter	*Ltn.*	*J26, 9*	28	30
Boddien, Hans-Helmut von	*Oblt.*	*J11, 59*	5	161
Boelcke, Oswald	*Hptm.*	*FA62, J2*	40	16
Boenigk, Oskar *Fr* von	*Hptm.*	*J4, 21, JGII*	26	35
Böhme, Erwin	*Oblt.*	*J2, 29, 2*	24	38
Bohnenkamp, Karl	*Vfw.*	*J22*	15	60
Böhning, Hans	*Ltn.*	*J36, 79, 32*	17	54
Bohny, Karl	*Ltn.*	*J17*	8	109
Bolle, Karl	*Rittm.*	*J28, 2*	36	19
Bona, Paul	*Vfw.*	*J1*	6	180
Bongartz, Heinrich	*Ltn.*	*J36*	33	23
Böning, Walter	*Ltn.*	*J19, 76*	17	54
Bönisch, Erich	*Ltn.z.S.*	*1 Seaplane Stn*	5	161
Borm, Gustav	*Uffz.*	*J1*	5	181
Bormann, Ernst	*Ltn.*	*J2*	16	58
Bowski, Hans	*Vfw.*	*J14, 51*	5	162
Boy, Herbert	*Ltn.*	*J14*	5	162
Brandenstein, Aloys *Fr* von	*Ltn.*	*J49*	8	109
Brandt, Franz	*Ltn.*	*J19, 27, 26*	10	107
Brauneck, Otto	*Ltn.*	*J25, 11*	10	92
Breiten-Landenberg, Otto von	*Ltn.*	*J9, 6, 11*	5	163
Brendle, Konrad	*Ltn.*	*J1, 45*	8	137
Bretschneider-Bodemer, Moritz-W	*Ltn.*	*J6*	6	180
Brünig, Helmut	*Ltn.*	*J32, 50, 41*	7	137
Büchner, Franz	*Ltn.*	*J9, 13*	40	17
Buckler, Julius	*Ltn.*	*J17*	36	19
Buddecke, Hans-Joachim	*Hptm.*	*FA23, 6, 5, J4, 30*	13	70
Buder, Erich	*Vfw.*	*J26, 84*	12	75
Buhl, Albin	*Obflgm.*	*SFS, MFJIV*	5	139
Bülow-Bothkamp, Harry von	*Ltn.*	*J36, 2*	6	139
Bülow-Bothkamp, Walter von	*Ltn.*	*FA22, 300, J18, 36*	28	29
Burckhardt, Friedrich-Karl	*Hptm.*	*J25*	5	163
Burkard, August	*Ltn.*	*J29*	6	140
Busse, Joachim von	*Ltn.*	*J3, 20*	11	84
Büttner, Siegfried	*Ltn.*	*J22, 61*	13	70
Camman, Theodor	*Oblt.*	*J2, 74*	12	75
Christ, Karl	*Ltn.*	*J28*	5	164
Christiansen, Friedrich	*Oblt.z.S.*	*SFL 1*	13	71
Classen, Fritz	*Vfw.*	*J26*	11	85
Collin, Dieter	*Ltn.*	*J2, 22, 56*	13	71
Creutzmann, Otto	*Ltn.*	*J20, 43, 46*	8	110
Croneiss, Theodor	*Oblt.*	*FA6*	5	164
Cymera, Wilhelm	*Off.Stv.*	*J1*	5	165
Dahlmann, Hermann	*Oblt.*	*J29, 26, JG III*	7	123
Dannhuber, Xaver	*Ltn.*	*J26, 79*	11	85
Degelow, Carl	*Ltn.*	*J7, 40*	30	27
Demisch, Martin	*Ltn.*	*J58*	10	93
Deilmann, Karl	*Ltn.*	*J6*	6	180
Delling, August	*Ltn.*	*J34*	5	165
Dietlen, Albert	*Ltn.*	*J23, 41, 58*	9	100
Dilthey, Helmut	*Ltn.*	*J27, 40*	7	123
Döbberke, Günther	*Ltn.*	*J45, 60*	8	110
Donhauser, Christian	*Ltn.*	*J17*	19	50
Döring, Kurt-Bertram von	*Rittm.*	*J4, 66, 1*	11	86
Dörr, Gustav	*Ltn.*	*J45*	35	20
Dossenbach, Albert	*Ltn.*	*FA22, J2, 36, 10*	15	60
Dostler, Eduard *Ritter* von	*Oblt.*	*J13, 34, 6*	26	35
Drekmann, Heinrich	*Ltn.*	*J26, 4*	11	86
Ehmann, Friedrich	*Vfw.*	*J47*	8	137
Ehmann, Gottfried	*Vfw.*	*S15*	12	76
Engelfried, Karl	*Flgmt.*	*SFS, MFJV*	6	180
Eschwege, Rudolf von	*Ltn.*	*FA66, 30*	20	46
Esswein, Otto	*Off.Stv.*	*J26*	12	76
Ewers, Walter	*Oblt.*	*J12, 77*	8	111

Name	Rank	Main units	Score	Page
Knappe, Herbert	Ltn.	J81	9	107
Koch, Emil	Ltn.	J32	7	128
Koepsch, Egon	Ltn.	J4	9	102
Kohlbach, Wilhelm	Ltn.	J50, 10	5	169
König, Erich	Ltn.	J2	6	148
Könnecke, Otto	Ltn.	J25, 5	35	21
Kopka, Johann	Off.Stv.	J62	5	181
Korff, Arthur	Ltn.	J60	8	116
Körner, Hans	Ltn.	J8, 19	7	129
Kosmahl, Fritz	Off.Stv.	FA22, 261, J26	9	102
Krebs, Fritz	Vfw.	J6	8	137
Kroll, Heinrich	Oblt.	J9, 24	33	23
Kuhn, Max	Vfw.	J21	12	80
Kühne, Wilhelm	Off.Stv.	J29, 18	7	129
Kummetz, Hans	Oblt.	J1	7	130
Kunz, Hermann	Ltn.	J7, 1F	6	148
Küppers, Kurt	Ltn.	J6, 48	6	149
Kypke, Walter	Ltn.	Kek A, J41, 47	9	103
Lange, Helmut	Ltn.	J26	9	103
Laumann, Arthur	Ltn.	J66, 10	28	31
Leffers, Gustav	Ltn.	Kek B, J1	9	104
Lenz, Alfred	Ltn.	FA61, J14, 22	6	149
Leptien, Hermann	Ltn.	J21, 63	7	130
Leusch, Wilhelm	Ltn.	J19	5	170
Lindenberger, Alfred	Ltn.	FA234, J2	12	80
Loerzer, Bruno	Hptm.	Kek J, J26, JGIII	44	15
Loerzer, Fritz	Ltn.	J5, 26	11	88
Löffler, Otto	Ltn.	J2	15	63
Lorenz, Heinrich	Oblt.	J33	5	170
Lotz, Paul	Ltn.	J7, 44	10	94
Löwenhardt, Erich	Oblt.	J10	54	13
Luer, Ludwig	Ltn.	J27, 62	6	150
Lux, Albert	Vfw.	J27	8	131
Mahn, Herbert	Ltn.	J72	9	104
Mai, Josef	Ltn.	J5	30	27
Mallinckrodt, Friedrich	Ltn.	Kek S, J6, 20	6	150
Manschott, Friedrich	Vfw.	J7	12	81
Marwede, Hans	Unt.	J67	5	171
Marwitz, Hans-Georg von der	Oblt.	J30	15	64
Matthaei, Rudolf	Ltn.	J21, 5, 46	10	95
Maushake, Heinrich	Ltn.	J4	6	151
Meinecke, Emil	Ltn.	FA6	6	151
Menckhoff, Carl	Oblt.	J3, 72	39	18
Mendel, Karl	Ltn.	J15, 18	7	131
Mesch, Christian	Vfw.	J26	13	90
Mettlich, Konrad	Oblt.z.S.	MFJI, J8	6	180
Meyer, Erich	Unt.	J45	5	181
Meyer, Georg	Ltn.	J22, 7, 37	24	38
Meyer, Karl	Obflugm.	SFS, MFJI	8	116
Mohnicke, Eberhardt	Ltn.	J11	9	105
Mohr, Alfred	Ltn.	J3	6	180
Monnington, Kurt	Ltn.	J15, 18	8	117
Müller, Hans	Ltn.	J12, 15, 18	12	81
Müller, Hans Karl	Ltn.	Kek A, J5	9	105
Müller, Max Ritter von	Ltn.	J2, 28	36	20
Mulzer, Max Ritter von	Ltn.	FA62, 32 Kek N & B	10	95
Nagler, Alfons	Vfw.	J74, 81	10	96
Nathanael, Edmund	Off.Stv.	J5	15	64
Näther, Max	Ltn.	J62	26	36
Neckel, Ulrich	Ltn.	J12, 13, 19, 6	30	28
Neuenhofen, Wilhelm	Ltn.	J27	15	65
Niederhoff, Alfred	Ltn.	J20, 11	7	132
Niethammer, Werner	Ltn.	J13	6	152
Noltenius, Friedrich	Ltn.	J27, 6, 11	21	44
Nülle, Hans	Vfw.	J39	11	90
Oberländer, Hans	Ltn.	J30	6	180
Odebrett, Karl	Ltn.	J16, 42	21	45
Osten, Hans-Georg v. der	Ltn.	J11, 4	5	171
Osterkamp, Theodor	Ltn.d.MA	MFJI, II	32	25
Osterroht, Paul	Hptm.	J12	7	132
Otto, Rudolf	Ltn.	J74, 68	6	180
Parschau, Otto	Ltn.	FA42, 261, KG1, FA32	8	117
Pech, Karl	Vfw.	J29	9	107
Pfieffer, Hermann	Ltn.	AOK3, J9	11	88
Piechulek, Franz	Ltn.	J41, J56	14	68
Pippart, Hans	Ltn.	K1, J13, 19	22	40
Plange, Richard	Ltn.	J2, 36	7	133
Plauth, Karl	Ltn.	J20, 51	17	56
Poeschke, Friedrich	Vfw.	J53	8	137
Poss, Reinhold	Ltn.z.S.	SFS, MFJIV	11	89
Pressentin von Rautter, Viktor von	Ltn.	J4	15	65
Preuss, Werner	Ltn.	J66	22	41
Pütter, Fritz	Ltn.	J9, 68	25	37
Pütz, Johann	Vfw.	J23, 34	7	133
Quandt, Theodor	Ltn.	J36	15	66
Raesch, Josef	Ltn.	J43	7	134
Rahn, Arthur	Ltn.	J19, 15	6	152
Ray, Franz	Ltn.	J1, 28, 49	17	56
Reimann, Leopold	Off.Stv.	J1, J2	5	172
Reimer, Claus	Ltn.	J26	8	118
Reinhard, Wilhelm	Hptm.	J11, J6, JGI	20	49
Richthofen, Lothar Fr von	Oblt.	J11	40	17

Name	*Rank*	*Main units*	*Score*	*Page*
Wenzl, Richard	*Ltn.*	*J31, 6*	12	83
Werner, Johannes	*Ltn.*	*J14*	7	136
Westphal, Siegfried	*Uffz.*	*J29*	6	181
Wiehle, Ernst	*Vfw.*	*S3, J43*	6	159
Wiessner, Ernst	*Ltn.*	*J18*	5	179
Windisch, Rudolf	*Ltn.*	*J32, 66*	22	42
Wintgens, Kurt	*Ltn.*	*FA48, 6, Kek V, J1*	19	51
Wissemann, Kurt	*Ltn.*	*J3*	5	179
Wolff, Hans-Joachim	*Ltn.*	*J11*	10	99
Wolff, Kurt	*Ltn.*	*J11, 29*	33	24
Wüsthoff, Kurt	*Ltn.*	*J4, 15*	27	34
Zander, Martin	*Hptm.*	*Kek N, J1*	5	180
Zenses, Alexander	*Vzflgm.*	*MFJII*	18	53
Zorn, Wilhelm	*Unt.*	*J60*	8	137

NOTES

NOTES

NOTES

NOTES

Also from the Publisher

WORLD WAR I GERMAN AVIATORS

THE SANKE CARDS

Charles Woolley

For the first time in eighty-five years, the famous Sanke postcards of World War I Imperial German Aviators have been reproduced. Clear, large scale copies of all known and highly collectible Sanke personality photo post-cards produced during World War I are now available, carefully replicated, and included all under one cover in this deluxe volume. Over 270 different cards of 132 individual aviators are included in this ground-breaking edition. Boelcke, Immelmann, the Richthofen brothers, Udet, and Göring are just a few of the famed aces and *Pour le Mérite* flyers photographed by *Postkarten-Vertrieb Willi Sanke*. Each postcard is given full page coverage, accompanied by a brief history of each man, together with his victories and highest attained award. A bibliography is included for the reader desiring to further research the lives of Germany's unusual, heralded, and greatest heroes of the First War. This book is a must for the student of uniforms as it depicts the amazing variety worn by the flyers of the *Luftstreitkräfte* 1914-1918, and also shows them wearing the multitude of awards and decorations presented to Germany's airborne heroes. Both private and museum collections have been combed to provide this book with the finest possible coverage ever compiled on the work of Willi Sanke and his talented photographers. This is an excellent companion volume to Schiffer's recently published *Aviation Awards of Imperial Germany, Volume VII* by the late Neal O'Connor.

Size: 8 1/2"x11" • over 300 b/w photographs • 272 pages

ISBN: 0-7643-1879-9 • hard cover • $69.95

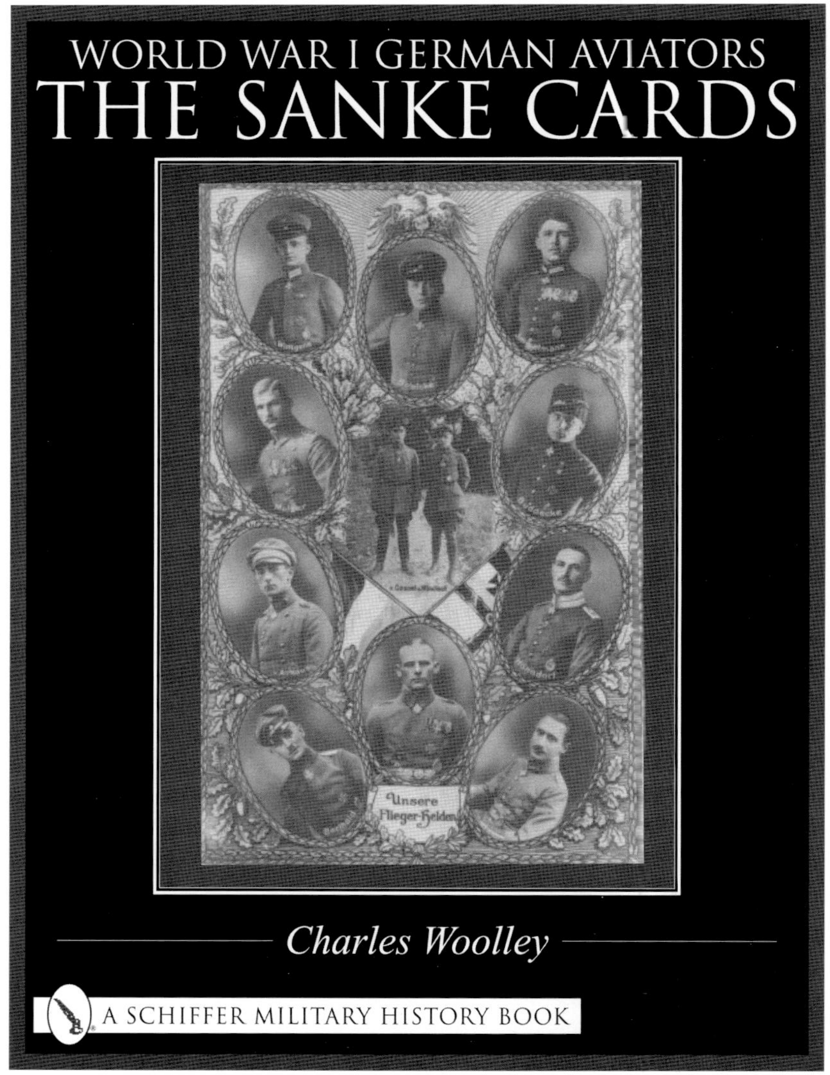